D1112870

MATHENAUTS

Books by Rudy Rucker

Geometry, Relativity and the Fourth Dimension
Spacetime Donuts
White Light, or, What Is Cantor's Continuum Problem?
Software
Infinity and the Mind: The Science and Philosophy of the Infinite
The Fifty-seventh Franz Kafka
The Sex Sphere
Light Fuse and Get Away
The Fourth Dimension: Toward a Geometry of Higher Reality
Master of Space and Time
The Secret of Life
Mind Tools: The Five Levels of Mathematical Reality
Wetware

MATHENAUTS

TALES OF MATHEMATICAL WONDER

EDITED BY

RUDY RUCKER

ARBOR HOUSE NEW YORK

Published in the United States of America by Arbor House Publishing Company and in Canada by Fitzhenry & Whiteside Ltd.

Manufactured in the United States of America

10 9 8 7 6 5 4 3 2 1

Library of Congress Cataloging-in-Publication Data

Mathenauts: tales of mathematical wonder.

1. Mathematicians—Fiction. 2. Mathematics—Fiction.
3. Science fiction, American. I. Rucker, Rudy v. B.
(Rudy von Bitter), 1946–
PS648.M33M38 1987 813'.0872'08 86-32184
ISBN 0-87795-891-2
ISBN 0-87795-890-4 (pbk.)

ACKNOWLEDGMENTS

"1 to 999" by Isaac Asimov. Copyright © 1981 by Montcalm Publishing Corporation. First published in *Twilight Zone* magazine. Reprinted by permission of the author.

"Four Brands of Impossible" by Norman Kagan. Copyright © 1964 by Mercury Press, Inc. First published in *The Magazine of Fantasy and Science Fiction*. Reprinted by permission of the author.

"Tangents" by Greg Bear. Copyright © 1986 by Omni Publications International, Ltd. First published by *Omni*. Reprinted by permission of the author.

"A New Golden Age" by Rudy Rucker. Copyright © by the Alumnae Association of Randolph-Macon Woman's College. First published in the R-MWC *Alumnae Bulletin*, Summer, 1981. Copyright © Rudy Rucker. Reprinted by permission of the author.

"Professor and Colonel" by Ruth Berman. Copyright © 1987 by Ruth Berman.

"The Maxwell Equations" by Anatoly Dnieprov, translated by Doris Johnson. Copyright © 1969 by New York University Press. First published in *Russian Science Fiction*, edited by Robert Magidoff. Reprinted by permission of New York University Press.

"Left or Right" by Martin Gardner. Copyright © 1951 by Esquire magazine. Copyright 1957 by Martin Gardner. Reprinted by permission of the author.

"Immune Dreams" by Ian Watson. Copyright © 1978 Ian Watson. First published in *Pulsar I* edited by George Hay, 1978. Reprinted by permission of the author.

"Forbidden Knowledge" by Kathryn Cramer. Copyright © 1987 by Kathryn Elizabeth Cramer.

"Gödel's Doom" by George Zebrowski. Copyright © 1985 by McGraw-Hill, Inc. First published in *Popular Computing*, February 1985. Reprinted by permission of the author and his agent, Joseph Elder Agency, 150 West 87th St., New York, NY 10024.

"The Tale of Happiton" by Douglas Hofstadter. Copyright © 1985 by Basic Books, Inc. First published in *Metamagical Themas* by Douglas Hofstadter. Reprinted by permission of the author.

"The Finagle Fiasco" by Don Sakers. Copyright © 1983 by Don Sakers. Reprinted by permission of the author.

"Convergent Series" by Larry Niven. Copyright © 1979 by Larry Niven. Reprinted by permission of the author.

"No-Sided Professor" by Martin Gardner. Copyright © 1947 by *Esquire* magazine. Copyright © renewed 1974 by Martin Gardner. Reprinted by permission of the author.

"Euclid Alone" by William F. Orr. Copyright © 1975 by Damon Knight. First published in *Orbit 16*. Reprinted by permission of Damon Knight.

"Love Comes to the Middleman" by Marc Laidlaw. Copyright © 1987 by Marc Laidlaw.

"Miss Mouse and the Fourth Dimension" by Robert Sheckley. Copyright © 1981 by Montcalm Publishing Corporation. First appeared in *Twilight Zone* magazine, December 1981. Reprinted by permission of the author's agent, Kirby McCauley.

"The Feeling of Power" by Isaac Asimov. Copyright © 1957 by Quinn Publishing Co., Inc. First published in *Galaxy* magazine. Copyright © renewed 1985 by Isaac Asimov. Reprinted by permission of the author.

"Cubeworld" by Henry H. Gross. Copyright © 1987 by Henry H. Gross.

"Schematic Man" by Frederik Pohl. Copyright © 1968 by HMH Publishing Co., Inc. First published in *Playboy* magazine. Reprinted by permission of the author.

"Time's Rub" by Gregory Benford. Copyright © 1984 by Abbenford Associates. Reprinted by permission of the author.

"Message Found in a Copy of *Flatland*" by Rudy Rucker. First published in *The 57th Franz Kafka* (Ace) by Rudy Rucker. Reprinted by permission of the author.

"The Mathenauts" by Norman Kagan. Copyright © 1965 by Mercury Press, Inc. First published in *The Magazine of Fantasy and Science Fiction*. Reprinted by permission of the author.

FOR CLIFTON FADIMAN

Contents

RUDY RUCKER

Introduction

Modulo a fuzz factor of two per modifier, *Mathenauts* is a collection of twenty science-fiction stories about math, all written since 1960. It is pleasant to have all these stories together in one place, initially to gorge upon, and then to compare and contrast.

The great thing about mathematical science fiction is that it gives the reader the weirdness of math without the work. You may be the kind of person who's interested in pursuing some of the ideas presented here—but you don't have to be. There's no homework, no problems, and no reading list. Just these crazy stories.

If you love math, you're probably not reading this, you've probably started right in on the stories. After all, there hasn't been a math SF antho since Clifton Fadiman's two collections, *Fantasia Mathematica* (1958) and *The Mathematical Magpie* (1962). Half a lifetime ago those books meant a lot to me, and to the people I hung around with. And now it's the late eighties, and it's been my happy lot to edit *Mathenauts: Tales of Mathematical Wonder.*

I could come on with a "math is fun" routine here, but why sell our squiddy muse short? Math is more than fun. Math is the study of abstract form, and everything has a form, so everything is math: number, space, logic, infinity, and information—also known as sensation, feeling, thinking, intuition, and communication.

I've reprinted two stories from the Fadiman volumes, but otherwise this is all new stuff, all post-1960. Not only are these stories new, these are all the new math SF stories I could find. Like some rare morel, our chosen delicacy appears but rarely—perhaps once or twice a year in fiction worldwide.

To the menu.

Who should lead off this collection if not Isaac Asimov? Who else but the indefatigable Good Doctor? "1 to 999" involves a cute puzzle, but the story's real charm is its masterly depiction of a popular type of fictional mathematician: the comically boorish genius.

A number of the other stories here are primarily sketches of mathematical types. Thus Ian Watson's "Immune Dreams" gives us a spaced-out self-destructive mathfreak, and William F. Orr's "Euclid Alone" shows us a moldering academic hack. Perhaps the most realistically presented mathematicians are those in Martin Gardner's "No-Sided Professor" and in Norman Kagan's stories.

Kagan's "Four Brands of Impossible" and "The Mathenauts" both came out in 1964. These are jittery, lacerating tales; the manic, hallucinated quality of "Four Brands of Impossible" reminds me of Thomas Pynchon's "Low-lands." It wouldn't be much of an exaggeration to say that I edited this book so that I could be sure to have copies of the Kagan stories on hand. In the event, Dave Hartwell at Arbor House suggested we name this antho after the Kagan story, and I think it a good idea. Who can forget a title like *Mathenauts*?

What ever became of Norman Kagan anyway? He left math for film and still lives in Manhattan. He's written a number of books on cinema, and is currently involved in putting together a TV science news magazine to be called "Spacetime Continuum News." When I pressed Norman for more math stories he confessed to having once written a mathematical *mystery* story called "The Venn Data Vendetta"—but he lost the only copy.

Some writers use science fiction as a sociological lens for looking at science. Others regard science fiction as a laboratory for thought experiments: what if. Questions: What if Earth were a cube? What if the fourth dimension were real? Each what-if world obeys its own inner logic, particularly if the initial dislocation is mathematical in nature. In the purest sort of "thought experiment" story the what-if acts as a new axiom that mathematico-logical reasoning exfoliates into a constellation of surprising yet inevitable consequences. Answers: If the world were a cube, then the corners would be like mountains—Henry Gross's "Cubeworld" demonstrates it. And if a four-dimensional creature were to enter our world, we'd see it as disconnected matter-globs—Greg Bear's "Tangents" shows us why.

Bear's story, which came out in 1985, is a slick and clever reworking of some of the classic four-dimensional notions that golden-age writers like Miles Breuer were fond of. Stylistically, "Tangents" has a cool, minimalist feel; spice is added by the use of a character based on the tragic British mathematician Alan Turing. Turing more or less invented the modern concept of the computer—theoretical computer science courses are largely about "Turing machines."

My own stories, "A New Golden Age" and "Message Found in a Copy of *Flatland*," are both what-ifs. What if there were a machine that could play math like music? What if Edwin Abbott's Flatland were located in the basement of a Pakistani restaurant? Perfectly reasonable questions, right? I wrote "A New Golden Age" after visiting the mathematical conference center in Oberwolfach, Germany.

Ruth Berman's "Professor and Colonel" is not really science fiction, although the machinery of H G Wells and of Michelson and Morley can be heard humming offstage. "Professor and Colonel" is primarily a character sketch of the most famous fictional mathematician of them all: Sherlock Holmes's archenemy, the criminal genius Professor James Moriarty. This is a lovely story, calm and lucid as a formal garden.

Anatoly Dnieprov's "The Maxwell Equations" was originally written in Russian, in 1960. Its science gimmick is the brainstim device that Phil Dick used to call a "Penfield mood organ." What makes the story memorable is its highly realistic use of math lingo. Dnieprov is serious about mathematics; his story has a continental aura of professionalism.

Martin Gardner's "Left or Right" has, to the best of my knowledge, never been reprinted anywhere. It first appeared in *Esquire* in the 1940s. Martin has resisted reprinting it all these years because he feared that the science had a basic flaw. He was worried that anything that's turned into its mirror image by a rotation through the fourth dimension automatically becomes antimatter—which could very much change the events of "Left or Right." But in reality we never actually *have* flipped something in 4D space, so there's no way of knowing; his story is safe.

Ian Watson's "Immune Dreams" is a wonderfully tetchy piece. No one would have written like this before the 1960s. "Immune

Dreams" also has the distinction of being the only story ever written about catastrophe theory. It's an odd coincidence that it contains the same cigarette-to-chalk shift as "Four Brands of Impossible." But maybe this isn't a coincidence so much as it is the efflorescing of an archetype. Chalk is, after all, a kind of drug for mathematicians.

Like "Professor and Colonel," "Cubeworld," and "Love Comes To The Middleman," Kathryn Cramer's "Forbidden Knowledge" is previously unpublished. Ms. Cramer is a mathematics student at Columbia. Rather than saying her story is *about* mathematics, one might almost say that, with its strange symmetries, it *is* mathematics.

As I was assembling *Mathenauts,* not all the people I told about it said, "Oh, boy." In fact some people said things like, "Are you kidding? *Math?* What kind of nut writes math stories? Ugh!" I'd like to sit those scoffers down and make them read George Zebrowski's "Gödel's Doom." Not so much because they would love the story, but rather because it would *confirm their worst suspicions*. George comes on like a math nut's math nut here—the story's a beaut. The "Gödel" of the title is Kurt Gödel, the Czech mathematician who spent most of his life at the Institute for Advanced Study in Princeton. In 1930, the year of the Great Depression, Gödel proved that no finite logical system can hold the infinite sea of mathematical truth. In a nutshell: there is no rational Secret of Life. But Zebrowski's story lets us wonder what if . . .

My two pop math books *Infinity and the Mind* and *Mind Tools* are good places to learn more about Gödel's theorem; other good places are Douglas Hofstadter's *Gödel, Escher, Bach* and *Metamagical Themas,* the latter of which yields "The Tale of Happiton." This is another story that really *is* mathematics. Doug goes wild with the number crunching and comes up with a convincing argument why each of us should spend at least fifteen minutes a day working for nuclear disarmament. I find this sortie from the ivory tower attractive and courageous.

Don Sakers's "Finagle Fiasco" is so silly that I had to love it. Question: does the story's Murphy Machine work or not? It's really a version of the famous Liar Paradox: Is the sentence "This sentence is not true" true? If it is, then it isn't; and if it isn't, then it is.

Larry Niven's "Convergent Series" is classic mathfun—an infinite regress. In a way, an infinite regress is a spatial version of a self-referential paradox. I love infinite regresses, and I was happy at the last minute to get another infinite-regress story: Marc Laidlaw's "Love Comes to the Middleman." Laidlaw pulls off a ramified doubly infinite regress. It's like being in Lilliput and Brobdignag at the same time. His writing has a nice freshness, a bit like Phil Dick in his happy pulp days. An odd but necessary feature of his what-if world is that *there is no way to get outdoors*. I guess all the rooms have light bulbs. Coming back to Larry's story, let yourself wonder how it would feel to be that devil, endlessly endlessly shifting down.

The opening sentence of Martin Gardner's "No-Sided Professor" is one of my favorite bits of stage setting. "Dolores—a tall, black-haired striptease at Chicago's Purple Hat Club—stood in the center of the dance floor and began the slow gyrations of her Cleopatra number, accompanied by soft Egyptian music from the Purple Hatters." I'd give anything to be there right now. Who says mathematicians don't know how to have a good time? "No-Sided Professor" dates back to 1946, when Martin was making a living as a freelance in Chicago. There's a couple of other Purple Hat Club stories, as a matter of fact, though they aren't SF. This is a wonderfully realized what-if story about topology, an interesting branch of math that started getting hot in the forties and is still going strong. As I write this intro, international topologists are debating whether or not the hypersphere is indeed the only simply connected 3D manifold. Lucky guys.

Anyone who has worked in academia will find William Orr's "Euclid Alone" depressingly realistic. It sometimes happens that older academics are more interested in protecting their turf than in encouraging the ongoing intellectual inquiry essential to the advance of science. "Euclid Alone" gives us a man like this who knows exactly what he is doing, and feels bad about it. Orr hones the conflict to a dissonant clashing edge. To fully grasp the story's ending, the reader needs to know that the title of the German paper means "A Geometrical Proof of the Inconsistency of the Axioms of Logic." This is a realistic concept because one of the spin-offs of Gödel's theorem is that we can never be totally *sure* that our system of logical reasoning is free of contradictions.

As a teenager, my favorite book of SF stories was Robert

Sheckley's *Untouched by Human Hands,* and my favorite novel was his *Immortality Incorporated.* Robert is still kicking around—he's a hard-boiled lifer writer. I got to meet him a few years ago when he showed up to live in our driveway for a week (he had a camper). I couldn't believe my good fortune. "Miss Mouse and the Fourth Dimension" is the only story ever written about Charles Hinton's cubes—an 1890s English mindtool for visualizing the fourth dimension. Speaking to me about the story, Robert summed up his researches into secret teachings with the phrase, "esoteric wisdom is what ends up on the remainder table." It's a great read.

Like "No-Sided Professor," Isaac Asimov's 1958 "A Feeling of Power" is a reprint from the Fadiman anthologies. This story is a classic because it does the impossible—it makes learning to multiply seem exciting. It even makes you want to learn how to extract square roots with pencil and paper! Well . . . maybe not *you,* but some of us, anyway. Simple, clear, thought-provoking stories like this have always been Isaac's forte.

Henry Gross's "Cubeworld" is a real find, perhaps the purest math SF story here. Not only is it a what-if, it asks a *new* what-if, an outrageous what-if that no one had ever thought of. Hank works at MIT, and has authored several novels.

One of modern SF's recurrent themes is the notion of coding the self as software. I'd thought it first appeared in John Varley's *Ophiuchi Hotline,* but here it is in a sixties story, Fredrik Pohl's "Schematic Man." Pohl does a wonderful job of explaining the concept of "mathematical modeling" here, and his main character is an unforgettable example of yet another mathtype: the self-effacingly brilliant seeker. I think the deep resonances of the story's theme relate to the fact that software replication is one of the few kinds of immortality we can count on. Every artist must ask: is it worth the price?

Gregory Benford's "Time's Rub" deals with a logical paradox known as Newcomb's problem, in which an individual is forced to make bets with an all-knowing croupier. The most science-fictional physics journal paper I know of is Benford, Book, and Newcomb, "The Tachyonic Anti-Telephone"—same Benford, same Newcomb. Besides making Newcomb's problem seem uncomfortably real, "Time's Rub" is also a fine example of Benford's cosmic mysticism, a monistic world view shared by many scientists.

When I wrote "Message Found in a Copy of *Flatland*," I was thinking of my friend Thomas Banchoff of Brown University, who was just then traveling to England in preparation for the centennial celebration of Edwin Abbott's *Flatland* (1884). Banchoff has created some remarkable computer-animated movies in which the screen effectively becomes a window looking into hyperspace. Others have taken up his techniques, and now there are a number of programs that will turn your computer monitor into a 4D scanner.

There are also some entirely different sorts of programs—known as cellular automaton programs—that use the computer viewscreen to explore even stranger mathematical landscapes. Last year I went to watch the people at the MIT Laboratory of Computer Science using the new CA scanners, and it struck me that, bit by bit, Kagan's "The Mathenauts" is becoming reality.

Clifton Fadiman managed to get two math SF anthologies out, and I hope that I'll be back in a few years with more of the same. There's a couple of classics I'd like to pick up next time; and there may be some more recent math stories that I've overlooked. But what I'd like to see most of all is a lot of brand-new math SF. No one's really done fractals yet, or cellular automata, or chaos theory, or information theory or . . .

What if?

Rudy Rucker
Department of Mathematics and Computer Science
San Jose State University
San Jose, CA 95192

MATHENAUTS

ISAAC ASIMOV

1 to 999

One can't help getting tired of Griswold at times. At least I can't.

I like him well enough. I can't help liking the old fraud, with his infinite capacity to hear in his sleep and his everlasting sipping at his scotch and soda, and his lies, and his scowls at us from under his enormous white eyebrows. But if I could catch him at his lies only once I think I would like him a lot better.

Of course, he might be telling the truth, but surely there can't be any one person in the world to whom so many impossible problems are posed. I don't believe it! I *don't* believe it.

I sat there that night in the Union Club with windy gusts of rain battering the windows now and then, and the traffic on Park Avenue rather muted, and I suppose my thoughts must have been spoken aloud.

At least, Jennings said, "What don't you believe?"

I was caught a little by surprise, but I jerked my thumb in Griswold's direction. "Him!" I said. "Him!"

I half-expected Griswold to growl back at me at that, but he seemed peacefully asleep between the wings of his tall armchair, his white moustache moving in and out to his regular breathing.

"Come on," said Baranov. "You enjoy listening to him."

"That's beside the point," I said. "Think of all those deathbed hints, for instance. Come on! How many times do people die and leave mysterious clues to their murderers? I don't think it has ever happened even once in real life, but it happens to Griswold all the time—according to Griswold. It's an insult to expect us to believe it."

It was at that point that Griswold opened one icy-blue eye and said, "The most remarkable deathbed clue I ever encountered had nothing to do with murder at all. It was a natural death and a deliberate joke of a sort, but I don't want to irritate you with the story." He opened his other eye and lifted his glass to his lips.

"Go ahead," said Jennings. "We are interested. We two, at least."

I was also, to be truthful—

What I am about to relate [said Griswold] involved no crime, no police, no spies, no secret agents. There was no reason why I should have known anything about it, but one of the senior men involved knew of my reputation. I can't imagine how such a thing comes about, since I never speak of the little things I have done, for I have better things to do than advertise my prowess. It's just that others talk, things get around, and any puzzle within a thousand miles is referred to me—which is the simple reason I encounter so many. [He glared at me.]

I was not made aware of the event till it was just about over so I must tell you most of the story as it was told to me, with appropriate condensation, of course, for I am not one to linger unnecessarily over details.

I will not name the institute at which it all took place or tell you where it is or when it happened. That would give you a chance to check my veracity and I consider it damned impertinent that any of you should feel it required checking or that you should go about snuffling after evidence.

In this unnamed institute there were people who dealt with the computerization of the human personality. What they wanted to do was to construct a program that would enable a computer to carry on a conversation that would be indistinguishable from that of a human being. Something like that has been done in the case of psychoanalytic double-talk, where a computer is designed to play the role of a Freudian who repeats his patient's remarks. That is trivial. What the institute was after was creative small talk, the swapping of ideas.

I was told that no one at the institute really expected to accomplish the task, but the mere attempt to do so was sure to uncover much of interest about the human mind, about human emotions and personality.

No one made much progress in this matter except for Horatio Trombone. Obviously, I have just invented that name and there is no use your trying to track it down.

Trombone had been able to make a computer do remarkable things, to respond in a fairly human fashion for a good length of time. No one would have mistaken it for a human being, of course, but Trombone did far better than anyone else had done, so there was considerable curiosity as to the nature of his program.

Trombone, however, would not divulge any information on the matter. He kept absolutely silent. He worked alone, without assistants or secretaries. He went so far as to burn all but the most essential records and keep those in a private safe. His intention, he said, was to keep matters strictly to himself until he himself was satisfied with what he had accomplished. He would then reveal all and accumulate the full credit and adulation that he was sure he deserved. One gathered he expected the Nobel Prize to begin with and to go straight uphill from there.

This struck others at the institute, you can well imagine, as eccentricity carried to the point of insanity, which may have been what it was. If he were mad, however, he was a mad genius and his superiors were reluctant to interfere with him. Not only did they feel that, left to himself, he might produce shattering scientific breakthroughs, but none of them had any hankering to go down in the science-history books as a villain.

Trombone's immediate superior, whom I will call Herbert Bassoon, argued with his difficult underling now and then. "Trombone," he would say, "if we could have a number of people combining mind and thought on this, progress would go faster."

"Nonsense," Trombone would say irascibly. "One intelligent person doesn't go faster just because twenty fools are nipping at his heels. You only have one decently intelligent person here, other than myself, and if I die before I'm done, he can carry on. I will leave him my records, but they will go only to him, and not until I die."

Trombone usually chuckled at such times, I was told, for he had a sense of humor as eccentric as his sense of privacy, and Bassoon told me that he had a premonition of just such trouble as eventually took place—though that may very well have been only hindsight.

The chance of Trombone's dying was, unfortunately, all too

good, for his heart was pumping on hope alone. There had been three heart attacks and there was a general opinion that the fourth would kill him. Nevertheless, though he was aware of the precarious thread by which his life hung, he would never name the one person he thought a worthy successor. Nor, by his actions, was it possible to judge who it might be. Trombone seemed to be amused at keeping the world in ignorance.

The fourth heart attack came while he was at work and it did kill him. He was alone at the time so there was no one to help him. It didn't kill him at once and he had time to feed instructions to his computer. At least, the computer produced a printout, which was found at the time his body was.

Trombone also left a will in the possession of his lawyer, who made its terms perfectly clear. The lawyer had the combination to the safe that held Trombone's records and that combination was to be released to no one but Trombone's chosen successor. The lawyer did not have the name of the successor, but the will stated that there would be an indication left behind. If people were too stupid to understand it—those were the words in the will—then, after the space of one week, all his records were to be destroyed.

Bassoon argued strenuously that the public good counted for more than Trombone's irrational orders, that his dead hand must not interfere with the advance of science. The lawyer, however, was adamant; long before the law could move, the records would be destroyed, since any legal action was to be at once met with destruction according to the terms of the will.

There was nothing to do but turn to the printout, and what it contained was a series of numerals: 1, 2, 3, 4, and so on, all the numerals up to 999. The series was carefully scanned. There was not one numeral missing, not one out of place; the full list, 1 to 999.

Bassoon pointed out that the instructions for such a printout were very simple, something Trombone might have done even at the point of death. Trombone might have intended something more complicated than a mere unbroken list of numerals, but had not had a chance to complete the instruction. Therefore, Bassoon said, the printout was not a true indication of what was in Trombone's mind and the will was invalid.

The lawyer shrugged that off. It was mere speculation, he said.

In the absence of evidence to the contrary, the printout had to be taken as it appeared, as saying exactly what Trombone wanted it to say.

Bassoon gathered his staff together for a meeting of minds. There were twenty men and women, any one of whom could, conceivably, have carried on Trombone's work. Every one of them would have longed for the chance, but no one of them could advance any logical bit of evidence that he or she was the one Trombone thought to be the "one decently intelligent person" among them. At least, not one could convince any of the others that he or she was the person.

Nor could a single one see any connection between that dull list of numerals and any person in the institute. I imagine some invented theories, but none were convincing to them all generally, and certainly none were convincing to the lawyer or moved him in any way.

Bassoon was slowly going mad. On the last day of the grace period, when he was no nearer a solution than at the start, he turned to me. I received his call at a time when I was very busy, but I knew Bassoon slightly, and I have always found it difficult to refuse help—especially to someone who sounded as desperate as he did.

We met in his office and he looked wretched. He told me the whole story and said when he had finished, "It is maddening to have what may be an enormous advance in that most difficult of subjects—the working of the human mind—come to nothing because of a half-mad eccentric, a stubborn robot of a lawyer, and a silly piece of paper. Yet I can make nothing of it."

I said, "Can it be a mistake to concentrate on the numerals? Is there anything unusual about the paper itself?"

"I swear to you, no," he said energetically. "It was ordinary paper without a mark on it except for the numerals from 1 to 999. We've done everything except subject it to neutron activation analysis, and I think I'd do that if I thought it would help. If you think I ought to, I will, but surely you can do better than that. Come on, Griswold; you have the reputation of being able to solve any puzzle."

I don't know where he got that notion. I never discuss such things myself.

I said, "There isn't much time—"

"I know," he said, "but I'll show you the paper; I'll introduce you to all the people who might be involved. I'll give you any information you need, any help you want—but we only have seven hours."

"Well," I said, "we may only need seven seconds. I don't know the names of the twenty people who might qualify as Trombone's successor, but if one of them has a rather unusual first name I have in mind—though it might conceivably be a surname—then I should say that is the person you are looking for."

I told him the name I had in mind and he jumped. It was unusual and one of the people at the institute did bear it. Even the lawyer admitted that person must be the intended successor, when I explained my reasoning, so the records were passed over.

However, I don't believe anything much has come of the research after all, unfortunately. In any case, that's the story.

"No, it isn't," I exploded. "What was the name you suggested and how did you get it out of a list of numerals from 1 to 999?"

Griswold, who had returned placidly to his drink, looked up sharply. "I can't believe you don't see it," he said. "The numerals went from 1 to 999 without missing a numeral and then stopped. I asked myself what the numerals from 1 to 999 inclusive had in common that numerals higher still, say 1,000, do not have, and how that can have anything to do with some one particular person.

"As written, I saw nothing, but suppose all those numerals were written out as English words: one, two, three, four, and so on, all the way up to nine hundred ninety-nine. That list of numbers is constructed of letters, but not of all twenty-six letters. Some letters are not to be found in the words for the first 999 numbers; letters such as 'a,' 'b,' 'c,' 'j,' 'k,' 'm,' 'p.'

"The most remarkable of these is 'a.' It is the third most commonly used letter in the English language with only 'e' and 't' ahead of it, yet you may go through all the numbers—one, fifty-three, seven hundred eighty-one, and you will not find a single 'a.' Once you pass 999, however, it breaks down. The number 1,000, 'one thousand,' has an 'a' but not any of the other missing letters. It seems quite clear, then, that the message hidden behind the list is simply the absence of 'a.' What's so difficult about that?"

I said, angrily, "That's just nonsense. Even if we were to admit that the message was the absence of 'a,' what would that mean as far as the successor was concerned? A name that lacked any 'a'?"

Griswold gave me a withering look. "I thought there would be several names like that and there were. But I also thought that some one person might have the name 'Noah,' which is pretty close to 'no a,' and one of them did. How much simpler can it be?"

*NORMAN KAGAN*______.

Four Brands of Impossible

"That concludes the Travis–Waldinger Theorem," said Professor Greenfield. "As you can see, it's really quite trivial."

"Then why did it take people to prove it?" piped up one teenage hotshot.

The bell cut off Greenfield's reply, and most of the class bolted. All the mathematics people at my school are bad—Greenfield the geometer was a mild case. His motto was: "If you can visualize it, it isn't geometry!" Which is not so bad compared to my other course, where rule one was: "If it seems to make sense, then it's not mathematical logic!"

Which reminded me I still had to find out about my grade in that subject, along with about aleph-sub-aleph other things—most important of which was securing a nice fat student-trainee job for the summer that was fast approaching. I elbowed my way past a couple of teenage hotshots, and then I was in the open air.

I decided I'd check out my marks later—the IBM-ed grades would be posted on the "wailing wall" all summer. Right now I'd leg it over to the Multiversity Placement Service. I could study for Greenfield's final this afternoon.

I walked across the campus slowly, checking myself out. "Do not judge according to appearance." Try telling that to some of these megabuck research corporations! I mussed my hair and put three more pencils in my breast pocket, and decided not to wear my glasses at the interview. It's amazing how much easier it is to lie to someone you can't really see.

"Hey, Zirkle—Perry—wait half a mo'!"

Harry Mandel hailed me from the psychology library. I grinned and waited for him to join me.

Harry is a swell guy, and besides he's a psychology major, not a math competitor. He joined me, puffing, a moment later. "Summer job hunting?"

"Yeah. I've got a couple of interviews arranged: Serendipity, Inc., and the Virgin Research Corporation."

"Me, too," said Harry. He gestured at my tousled hair. "Getting ready? Physical appearance is very important, you know." The short, pudgy psychology major pumped his legs to keep up with me.

It was a warm, comfortable day on the Multiversity campus. The long rows of wooden chairs were already set up for graduation, and here and there was a girl in long hair and Levis, or a bearded boy with a guitar. Early summer-session people. Lazy jerks.

"What's the word on the companies?"

Mandel wrinkled up his forehead. "Personnel men—not technical people. So if you've got the grades, go all out—anything to avoid the paper barrier."

"Any specific suggestions?"

"Mmmm—well, Fester pulled a full-scale epileptic fit—but then he's nearly a five point. If you're just bright, a few eccentricities ought to do it. I'm trying my bug-on-the-walls gambit."

"You mean the one where you pretend there's a bug that crawls all over the walls behind the interviewer, and you follow it with your eyes."

"No—that was last year. In this one I sort of scrunch up in the chair, cowering—give 'em the impression I can't stand confined spaces, need lots of room—like, say, New Mexico or Arizona. I'm sick of this East Coast weather, and the Virgin Research Corporation has labs in New Mexico."

"That's for me, too. I'll see what I can think up."

In the 1980s it's practically impossible to get a summer job in the sciences—not that the big science and engineering corporations don't want you. They do. But to apply, you've got to submit about a ton of paper work—eight commendations, four transcripts, character references, handwriting samples, personality profiles, certificates and forms and diplomas. Who has the energy?

Science and engineering majors, however, have worked out a

swell dodge—we just pretend we're a little bit nuts. The big company personnel departments are endlessly amused by the antics of their nutty research wizards. With the squeeze on for technical people, it's easy to fake looniness well enough for the personnel men to see that cause follows from effect, in that wonderful way of theirs, and conclude we're the brilliant boys they're looking for.

Or maybe they just get a kick out of watching us degrade ourselves in front of them.

I had a couple of swell dodges I'd worked out from one of my professors—through the length of the interview I'd keep pulling a piece of chalk from my pocket, sticking it into my mouth, then spitting it out and muttering, "Simply must give up smoking!" For dubious types, I'd offer my *pièce de résistance;* all through the talk I would gesture and wave my arms, seeming to shape the very job concept out of the air. Then, when the interview had reached a critical juncture, I'd pause, drop to the floor, and lie on my back staring at the empty spaces I'd been manipulating. As the interviewer came round the desk, I'd cry out in annoyance, "Simply must look at this from a new point of view!" It worked like a charm.

Except for this time. Not that it didn't; I just never got the chance to use it. On this job, all the craziness came at the end.

The interviewer for the Virgin Research Corporation was a big blond crew-cut man with terribly stained teeth and a sadist's smile. He reminded me of one of my philosophy professors. He was talking philo at me too, about half a second after I sat down in the little interview cubicle.

"Glad to meet you, Mr. Zirkle—you're a mathematics major, by your application. Is that right?"

I nodded.

"Sit down, sit down," he said, gesturing. "Now, before I begin to ask about you, I'd like to tell you a little about the activities of the Virgin Research Corporation—Mama, as we call her around the shop. Our organization is concerned with the three aspects of pure research, what we like to call 'The three brands of impossible.'"

I nodded at this. Harry Mandel's eyes had been shining when he'd passed me outside the room, but he hadn't time to whisper more than a fiery; "Grab it, fellow!" to me before I was ushered in. I hunched forward and began to listen.

"If we ignore subjective problems—what the Kansas farmer said when he saw his first kangaroo—we might analyze the concept of the impossible as follows."

He pulled out a diagram, and his finger danced down it as he continued to speak.

"First, there is the 'technically impossible'—things that are not possible in practice, though there's no real reason why they can't be done. Things like putting the toothpaste back in the toothpaste tube, or sending an astronaut to Saturn—such things aren't practical at the moment, I think!" he said, smiling briefly.

"Then there is the notion of the 'scientifically impossible'—traveling faster than light, or building a perpetual motion machine. These are not possible at all—within the limitations of what we know about the universe. But you'll recall, a heavier-than-air flying machine was a 'scientific impossibility' a century ago.

"These two categories have merged somewhat in the twentieth century, though the distinction is clear enough. In the first case, the 'technically impossible,' theory allows you to do the impossible—you just haven't the techniques. In the second instance, the 'scientifically impossible,' you've got no *theoretical* justification for what you want to do. But in both cases, men have 'done the impossible'—either developed new techniques or found the flaws and limitations in the theories.

"But there is a third category of the impossible, one ignored by even the most farsighted researchers—the 'logically impossible'!" The interviewer's voice rose in triumph, and his other hand, which had remained in his pocket, furiously jangled his change.

I blinked at him. "But the logically impossible is—"

"I know, I know, I've listened to that stuff from our professional consultants," said the big blond man, suddenly impatient. "The logically impossible is part of an arbitrary system which would be destroyed by any attempt to—" he shrugged his shoulders in annoyance.

"Let me tell you," he cried, "that the Virgin Research Corporation has investigated the problem and decided otherwise. Our experts have—and they are some of the best men in the field, much better than any jerky Ivy League Multiversity can afford—our experts are convinced that such notions as the 'round-square' are meaningful, and what's more, are of potentially great military value!"

His eyes were crazed. "'When the battle's lost and won,' indeed," he murmured in a low, sinister voice. He smiled at me coldly, and the rotten stains on his teeth stood out like the craters of the moon.

"We're using a two-pronged approach—psychology and mathematical logic. We've had no trouble recruiting psychology majors," he continued in a normal tone, "but most of the students in the mathematics department weren't interested—or they've got to spend the summer with their families at home or away."

It was my turn to grin. That's what they got for trying to interest any of the teenage hotshots. But I wasn't afraid to broaden my mental horizons, I was willing to wrestle with the impossible, I was brave enough to face the unknown. My smile widened, and then my face grew serious as I took up the challenge.

"How much?"

"Two-fifty a week, recommendations, room and board and a motor scooter, and free transportation in a G.E.M. cruiser to and from the New Mexico labs," said the blond man.

"Well—"

"With your record, you should jump at the chance," he said. "I've seen your transcripts, son." He began jangling his change again.

That "son" decided me. "Hmmmm—"

"All right, all right, we'll talk it over," he said, a little sharply. In science and mathematics, all the old guys are scared of all the young guys. You do your best work when you are young, and everyone's scared of being "burned out at thirty." Just like I'd love to line my teenage competitors up against a wall and plug 'em, I could see he was scared of me.

He pulled out my preliminary transcripts and applications and began thumbing them. I slipped off my glasses and licked my lips.

Later I found out I got A-minus on the mathematical logic final. I should've asked for fifty more than the three-twenty-five I got out of that terrified jerk.

The G.E.M. *Ruby* thundered west, the ground effect keeping it a dozen feet above the earth. The machine soared along, impossibly graceful, as night mastered day on the American Great Plains.

I peered out the big picture window, fully relaxed for the first time in many weeks. The orientation and information kit lay ignored across my knees. I'd look at it later. No more lab reports, no more little phrases like "I'll leave that as an exercise," that meant a dozen hours of skull sweat; no more "I'm sorry, but some pre-med sliced those pages out of the book you wanted with a razor blade last term." At the moment I didn't care if the directors of the Virgin Research Corporation had cerebrum, cerebellum, and medulla in their brainpans, or scrambled eggs. By Napier's bones, I'd escaped!

Someone was struggling up the aisle against the pressure of the *Ruby's* acceleration (we'd just pulled out of Ann Arbor). With a gasp he collapsed into the acceleration chair beside me. "Greetings!" I murmured. "You one of Mama's boys?" About half the people on the *Ruby* were working for V.R.C. It's only these tremendous mysterious corporations that can afford intercontinental jet flights and G.E.M.s and—and pure mathematicians, thank goodness!

"Hello, yes," said my companion. He was a skinny, baffled-looking fellow about my own age. His very pale face said, "Yourself?"

"Perry Zirkle—I'm in the numbers racket—uh, I'm a pure mathematician."

"Uh, Richard Colby—microminiaturization and electronics—I'm a grad student at Michigan Multi. If you can see it, then it's too big. My motto." Colby's face brightened and he grinned. *His* teeth were okay. "Say, I've got those books—you must be on the logical impossibility research the same as I am—Project Round-Square!"

I nodded and smirked at the books. "I suppose so—though from what I've been taught, I doubt if the project will last very long."

Colby settled himself and relaxed. "How so?" he asked. He didn't look like a monomaniacal studier—just an electron pusher in his twenties. He wasn't one of these kid geniuses, either, and I was rested and relaxed. So naturally, my mouth got the better of me.

"Just on the face of it—," I argued calmly. "Paradoxes and self-contradictions are interesting, and they attract attention to ideas, but by their very nature—" I found myself unable to continue.

"Maybe," Colby said. "But maybe you're just looking at the problem the wrong way—the fellow that interviewed me kept talking about 'thinking in other categories.'"

I paused. "Oh, I know what he meant," I said, and laughed. "He was trying to tell you not to argue, not at two-fifty a week."

"Two-twenty-five," he murmured.

The electronics expert hesitated, and then looked at me oddly. "I don't know about you," he muttered, "but I consider it an honor and a pleasure to be able to do some 'pure' research. There's little enough of it in electronics these days—the whole subject has about one real scientist to a hundred engineers." His eyes were hooded. In the dimly lit passenger compartment of the G.E.M., his face was dark and brooding. He licked his lips and went on, talking to himself as much as he was to me.

"It's enough to make you go into industry. Take my own school, the Michigan Multiversity. Did you know we have a top secret Congressional Project to automate the presidency? Fact. The chairman of the Department of Cybernetics told me the system philosophy behind it: "Roosevelt showed that someone could be president as long as he liked. Truman proved that anyone could be president. Eisenhower demonstrated that you don't really need a president. And Kennedy was further proof that it's dangerous to *be* a human president. So we're working out a way to automate the office." He grinned, and I laughed in response.

I reached down into my fagbag and pulled out a bottle. His eyes went wide for a moment, but I passed it to him. He took a slug, and the evening was on its way.

Colby turned out to be all right. I told him Smith's remark about how engineers are sloppy when they call "characteristic values" "eigenvalues," because "eigenvalue" isn't good English. He came back with the one about the sequence that you had to prove converged, but that all the students demonstrated diverged. The professor's masterful reply was "It converges *slowly*."

The ground-effect machine rushed on through the midwestern night, a foot or two above the earth, supported by a flaring cushion of air. Presently its path curved south. The pilot-driver was steering by radar beacon and navigation satellite, towns and buildings signifying no more than treacherous shoals and reefs to a sailor. The craft was flying over ground that had never, and might now be never, touched by wheels or feet. Over these wastes we plunged southwest.

Dick Colby couldn't hold it very well, or maybe he was tired. In any event the fellow was soon sleeping peacefully beside me. I let him be and stared at the scenery.

These fellows that believe the "pures vs. applieds" battle really amuse me. Actually, science and scientists are just like anything else in this rotten world, just as corrupt. I've heard stories of research men during the great "Space Flight Bubble" that would trade jobs a dozen times in a year, doubling their salary each time. And these stories about advertising men that run off with the best accounts and start their own agencies? Nothing to the technical men that impress the Pentagon and get the generals to finance them in their own electronics company. Though I don't feel much sympathy for the big firms. Anyone that builds H-bombs and missiles and lets someone else decide what to do with them—people like that deserve everything they get!

What was wrong with me lately? I still loved to work and study, to cram till one, then feel the high tension as the papers were handed out the next day. The gong that announced the start of the test always reminded me of the one on the old TV show, "Beat the Clock." And there was nothing like the feeling in front of the posted grades when I saw the shocked faces of the youngsters I'd beaten out. Tough luck, kid! Better switch to art history!

I lit a cigarette and leaned back. Well, right or wrong, this stuff would be fun. Science always is. I love to be totally absorbed in something new and strange. It's so much better than just sitting around doing nothing, or dull routine stuff. Frankly, I don't see how the hundred million unemployed can take it. My mood when I'm idle is usually a murderous rage at the kids who're going to parties and dances and junk like that. Not that that stuff is really *interesting,* like a problem in Greenfield space. But at least it's something, compared to sitting all alone with nothing to think about but myself. Frankly, I love really tough problems, the kind you have to think about *all the time.*

Dawn was peeping up over the horizon. I settled myself in my own acceleration chair and tried to snatch a little sleep. My own watch said that in a few hours we'd arrive at the immense desert reservation which held the Virgin Research Corporation, summer student trainees for the enigmatic Project Round-Square.

"These are your quarters, Mr. Colby and Mr. Zirkle," said the blond girl. She was worth a second glance, being the possessor of a fine body, though a little bow-legged. ("Pleasure bent," Colby murmured.) Still, a very nice body.

Colby dumped his junk on the bed and began opening drawers in the dresser and putting it away. I stood still and read the information sheet we'd been given on arrival. It said I had to report at the Computer Center as soon as convenient. I put my own bags in the closet and went.

Outside, the desert sunlight was quite bearable, since it was only a few hours after sunup. I walked across the compound, guided by a map on the fact sheet.

The living quarters were good: simple ranch-style stuff with desks, and bookshelves duplicated in each room. This was no resort, but the place was clean and kept up, without the bleakness of a straight government installation. The labs and auxiliary buildings were spread out over the desert, the whole business enclosed by a security frontier. This made internal security checks unnecessary, and there were none.

People dressed informally: chinos, dungarees, western boots and flannel shirts. A pleasant change after school, where most everyone was formal most of the time—except for the technical students.

Of course they made us pay for it. All the co-eds are hot for someone they can discuss the Great Books with, not some barbarian science or engineering major with a slide rule swinging from his belt. I've seen these Zenish girls, with their long hair and thongs and SANE buttons, wild for motorcyclers and African exchange students. Rotten snobs! Though I've got to admit that some of my friends in the engineering school depend more on force than persusion for their pleasure. Ha-ha!

The Computer Center was mostly underground, to make temperature regulation easier. These big machines really heat up. I know that back at the Arthur Regleihofp Computing Center, at my own multiversity, they have an enórmous air conditioning plant through all the machine rooms, with dozens of recording thermometers. If the temperature in the labs goes above a certain point, the electrical power to the computers is shut off. Otherwise you have something called a computer explosion, which no one at the labs wants to talk about. All I know is that during the summer the machine rooms are the best place to relax, because they're so cool. And I can always scare the teenage geniuses who run them into letting me rubberneck.

So I reported to the Computer Center, and found my wonderful creative position from which I could challenge the unknown—the programming saddle of an obsolete IBM aleph-sub-zero—a jazzed-up Turing engine. The same noble trade I'd learned six years before, as a youngster in the Science Honors Program at the multiversity.

An International Business Machines aleph-sub-zero tests a mathematical model against reality. The device begins grinding out deductions from the model, and checking them about facts about the phenomenon. If they check out, fine. If not, it begins to blink and tremble in agitation.

This one had a few peculiarities. The "mode" was about ten times as complex as normal, there were fifty more storage units—and the runs averaged less than ten seconds.

By the end of the day, I was bored, frustrated, and very disgusted. I could barely keep from grabbing my teenage assistant by his ankles, swinging him around in a heavy arc, and smashing out his smiling freckled face against the machine's one-to-ones. Rotten teenage competitor! Fortunately I ran into Harry Mandel directly afterwards, without having to look him up. At least that was something. One thing led to another, and two hours later, together with Richard Colby, the three of us were exchanging impressions.

"Oh, I suppose it's all I could expect," I told them disgustedly. "They're setting up odd sorts of logical-mathematical models—ones without the law of self contradiction, either A or not-A. Things like that. Then they run them through the aleph-sub-zero as a check. Only—" and I took a deep slug from my glass, "—none of them work."

Harry Mandel bobbled his head up and down enthusiastically, so that it seemed to flicker in the cool dim corner of the White Sands Bar. Harry has this habit of shaking his head in violent agreement, while his eyes grow larger and larger with each sentence you speak. It gives you the funny feeling that every word you say is confirming some incredible theory of his: that you're a Chinese Communist, or a paranoid schizophrenic, or an Arcturian spy. It's really quite frightening until you get used to it.

Also, his lips were trembling and his hands quivering. I knew the signs. Once he started talking, he'd never stop. So I gave the nod to Richie Colby instead.

The electronics expert looked up from his drink. "I don't know," he muttered. "I'm on the psychological-biological end, and so far I can't understand what's going on. They've got me working on topological neuronic maps—mapping the circuits of the brain. But for what, I don't know." He went back to the drink he was nursing.

I took a sip from my Coke. I don't drink more than I have to, and neither do most of my friends. In spite of all this talk about college students boozing it up, I'll be damned if I'll rot my brains, the brains that have to beat out all those teenage hotshots!

The White Sands Bar was a pretty good one, quiet with a kitchen. A while before we'd had a pizza, heavy with cheese and olive oil. It's funny how much time I spend in bars. Our civilization has wonderful extensive facilities for some things, fragmentary ones or none at all for others. It's perfectly clear how to fill out the forms and go to class and take exams and apply and student all the way to a Ph.D.—but how the hell do you have a good time? I heard they had to double the psychiatric service up at MIT. Sometimes I have crazy insane dreams of getting out of this whole mess, quitting. But where could I go, what would I do, who would be my friends? *Who would be my friends?* Anyway, bars are all right, and this bar, the White Sands Bar, was a pretty good one.

I took up another drink. Richard Colby was staring dumbly into his. "Okay, Harry," I said.

"To understand my end, you'll need to know what the universe is," said Mandel quickly and incoherently. "People ask: 'Why is the universe the way it is?' And Kant answered them back: 'Because the universe is a tango!'"

"Huh?"

"Don't you know what a tango is?" said Harry quizzically. "Why, even all my buddy-buddy psychology-major friends know *that*. You know—like this—" and he moved his hips suggestively. He grabbed up his sloe gin and Coke and finished it off in a single gulp. "Daiquiri!" he cried to the waitress. "Like this, boys," he moaned, beginning to sway again.

It took a little while to make our questions clear, but presently Harry was sketching on a napkin with his Mr. Peanut pen.

"Remember that proof in high school geometry—tenth-year mathematics to you, Perry—where you show that a line segment has only one perpendicular bisector. You strike arcs from the end points, and draw the line from one intersection to the other. But why should the arcs have any intersection? And why couldn't there be *two* lines that were straight and went through both intersection points? I bet you never thought of that!"

He looked up from his diagram defiantly.

Dick Colby blinked at him, his long face weary.

"I'll *tell* you why!" cried Harry Mandel, downing half his daiquiri. He put the glass down and spoke decisively. "Because the universe is a tango—we see it this way because *we have* to—we're built this way. Anything else would be a logical impossibility—a contradiction. We can't experience the world any other way. We see it this way because we're built a certain way, and the universe is built a certain way. Reality is the interaction of the two parts—and the universe is a tango!"

He tossed down the rest of the woman's drink and nodded powerfully. "Any one of my buddy-buddy psychology-major friends will tell you that!"

Colby and I nodded agreement. Mandel always was something of a nut. I never trust short guys—their mothers always tell them about Napoleon when they're little, and they always take it the wrong way.

Mandel was still gabbling. "But this doesn't mean we'll always have to look at things this way. We won't have to always think of a round-square as impossible. That's what my part of Project Round Square is all about, the part with my buddy-buddy psychology-major friends. We're going to change the music. We're going to give one partner dancing lessons!"

It was five weeks more before I learned that Mandel had in fact not been kidding around, nor really drunk at all. That was the essence of the psychology half of Project Round-Square. But a lot happened between that first night at the White Sands Bar and then.

For one thing, they closed down the mathematics-logic side of the installation. I had about a week more of that "Start Program!" Zip-pip-pip-pip-pip! "Clang! Clang! Clang! Discrepancy! Discrepancy!" nonsense, then two days of absolutely flawless correlations—as good as any of the test runs between economics and high school math, or advanced calculus and statics and dynamics. Whatever was being sent into the aleph-sub-zero, it was a perfect fit with the real world. The first day I was wildly enthusiastic, the second I was bewildered—maybe they were checking themselves? And the third day, I wasn't given any programs. The head of my section, a young man named Besser, showed up about an hour later and told me we were shutting down. I was to be reassigned.

"But why? The last two runs were perfect!"

"The last runs—" he began, then sighed. He looked more like a truck driver than a worker with the subtle squiggles of mathematical logic. "The last two runs were exercises in futility. You've had some undergraduate symbolic logic—you must have some idea of what we're trying to do."

I nodded.

"Well, rigorously speaking, the way to eliminate the notion of 'impossible' is to get rid of contradiction—get a sort of logic where you can have a round-square as a legitimate notion. Then you build a language with that logic. Understand me?"

"Uh-huh."

"Now, this seems—ahem—unlikely. If you've ever taken an introductory course in philosophy, there's always a kid who talks about there being 'some crazy kind of logic' where things could be red *and* blue, round *and* square." He shrugged in annoyance. "The professor can usually shut him up, and if he's persistent, embarrass him to death. Those kids are the sort that embarrass pretty easily."

I nodded. This guy knew something of college life.

"So that's what the meta-mathematicians and symbolic logicians upstairs have been working on. You see, while such things are silly to talk about here in the real world, you *can* have a logic without the 'not' operator. Such logics have been set up in the past—but they weren't very interesting, they weren't *rich,* fruitful in new ideas. But anyway, you can make such a thing, you can even build up to a mathematics from it, the way Russell and Whitehead built up numbers from logic in *Principia.* And you use your math to build a logic and a language—to describe the world. No 'not' means no opposites—which seems to mean no contradictions." He wiped his face and tried to look annoyed, but it was difficult. Good old air conditioning.

"Do you see?"

"I think so. Real world again—didn't match up."

"Kee-rect. Your math is no good for the real world. It's just wrong—like trying to navigate an ocean liner with plane geometry. Since the earth is round, it doesn't work out."

I nodded.

"I mean, it's *right*—it's *valid*—it just doesn't describe anything real," he corrected hastily. "Seems as if you *must* have contradiction."

"That's why the runs on the aleph-sub-zero were so short? The computer would spot a contradiction and start yapping. But how about the last two runs—perfect straight out. What was the matter with them?"

"Oh, those," he groaned. "Those were that jerk Kadison's idea. The exclusion approach."

"Go on."

"Well, you know there's another way to eliminate the notion of a contradiction. By exclusion."

"Elucidate."

"Think of it this way," said Besser. "You understand the notions of tall and short, and you know such things are relative. But if you decide that everyone under twenty feet tall was short—then you couldn't have a contradiction, a notion of 'short-tall.' Everyone would be short, and you'd eliminate one sort of contradiction. 'Tall-short' would mean the same as 'mimsey-short'—'nonsense word-short', or just 'short.' And you keep on going that way. This was Kadison's idea."

"It worked perfectly in the machine."

"Sure it did. And it's also perfectly useless. *All* gradations and comparisons drop out—and brother, you don't know how many there are. You know, most every quality has its opposite in *something* else. Even the notions of matter and empty space. You get nothing left—I mean *nothing*—the problem becomes trivial."

"The universe is an uncle," I said.

"Yeah, except for uncle say any other word. The universe becomes one solid, undescribable lump, with no qualities at all."

Also, Harry Mandel began to crack up. I didn't notice it while I was working—that aleph-sub-zero had some good problems—but when I was unassigned, the only real activity I had was going to the White Sands Bar. It was during the drinking sessions that his madness began to blossom.

Now of course I know all about the science of modern psychology—it's one reason I've remained so balanced and stable. Myself, I'm what is known as a shame personality. It's a matter of personal honor with me that I fight to the limit for the highest grades and the most scholarship. I'll beat 'em all out. That's me.

Dick Colby was clearly a guilt personality. He really believed all that guff about the scientist's world view, about the search for truth, about following an abstract pattern of behavior. Poor old Dick, he had an abstract moral code too, as I might have expected. Well, he had to follow the rules, and he might get kicked in the belly, but at least he was stable in his sad, picky way.

Now Harry Mandel was a fear personality. He tried to belong to some sort of whole, to link its destiny with his own. A gestalt. You find a lot like that: fraternity boys, soldiers, club members, athletes on a team. And of course the intellectuals—the literary group and Zenish girls and the interdependent independents.

Ain't social psychology great!

Anyway, Mandel's kind of twitch, the fear personality, is okay as long as he really has his buddies and believes it. If he doesn't have them, he wanders around until he can link up with a new bunch. If he doesn't *believe* he has them—look out! A fear personality with doubts about its gestalts can slide right over into paranoid schizophrenia.

It came out in funny ways, distorted, because Mandel was very intelligent, and the more intelligent, the more little links can begin to snap and break. One night in late August he came up with this:

"I mean, I have nothing against that particular minority group," he said loudly. "It's just that—well, look at it this way. The original members were selected on the basis of crude physical strength—the smart clever ones escaped the slavers. Then they were brought over here, and were slaves for several hundred years. Now it seems to me that if you have slaves, you're going to encourage breeding among the stupid and the strong. You don't want smart quick ones. As a matter of fact, the smart quick ones would try to escape, and would be shot. Or else, if they're clever, slip over the color line and intermarry.

"So you see, you've had forces at work for three hundred years that bred—I mean in terms of human genetics—for less intelligence. You do that for three centuries and it shows—as a matter of fact, it *does* show. In terms of modern science, they might even *be* inferior."

He was crazy, insane. For one thing, three hundred years isn't long enough to matter genetically for humans. For another, "the smart clever ones" *didn't* escape the slavers' roundups any more than the others did. Maybe if Mandel had been a slave owner he'd have tried to encourage only the stupid to breed, but such eugenically oriented thinking didn't exist in times past. As for shooting would-be escapees, you don't destroy valuable merchandise like that, you bring it back alive. [Anyway, you don't talk about races, you talk about human beings.] Mandel was rationalizing, justifying immoral attitudes on the basis of a "science" which really doesn't exist. As for "scientific morality," hell, science and morality are different, and by trying to base one on the other you are setting up for something like Hitler's "final solution."

Yet poor disturbed Mandel had *thought the theory up*. And for an instant, thinking of the exchange students and the Zenish girls back at the multiversity, my own brain had become enflamed. Science, reason, intellect—there are some things you mustn't think about. God help me.

A flash of disgust went through me. I never wanted to do any more calculations. I wanted to lie down with some pretty girl and make love to her and have her soothe me. I was too long alone. Help me! Then I squeezed those thoughts away.

I pushed away the thick silvery tin with the remains of the pizza. Delicious, too. Their cook was improving. Or maybe he could get real Mexican spices, not the stuff I used to settle for in Woolworth's.

I looked at Mandel across the table. His face was beginning to cave in a little, and his eyes looked tired. Anyway, we had to talk about something else. I decided to find out some things about my new assignment. And that was thing number three. That afternoon I'd be transferred to the psychology attack team-programming again, the Urbont matrices of neurological maps.

"How's the job, Harry?" I asked him. "How's the dancing lessons coming?"

Mandel looked up. He hadn't mentioned his work in a couple of weeks, not since we'd taken his little lecture on Kant and the tango as a rib.

"What about it?" he asked crisply. "I don't know very much, I just do whatever they tell me."

They—his buddy-buddy psychologist buddies? Mmmmm.

"Well, whatever they've been telling you, they'll soon be telling me," I said, nudging him in the shoulder. "I'll be figuring out your brain maps for you."

"Oh, yeah. They ought to start programming in two weeks, and then installation—"

"Hey, installation? Whoa? What are you talking about?"

Mandel's slumped body seemed to collapse some more. He was so far forward I could hardly see his face. Just a dark form against the well-lit rest of the bar. Cigarette smoke hung in the air, and a dozen technicians were seated on the high stools. Over in another corner, two disgruntled physicians were playing nim. A couple of the girls from the clerical pool were having heroes and Cokes at another table, blond and brownette in brief desert costumes.

"Sensory enervation," said Mandel in a dead voice. "What else did you think?"

He blinked mildly and slugged down the rest of his horse's neck. "Rum and Coke!" he shouted to the barman. "I really hate to drink," he confided sullenly. "But at least I can do it alone, without my psychology-major buddy-buddies." Richard Colby gave me a funny look and we both leaned forward and really began to listen.

In the 1950s, the psychologists of McGill University had commenced an interesting sequence of experiments in connection with the U.S. manned space-flight program—an early phase of the "Space Flight Bubble." A space traveler confined to his space capsule would be in a state of extreme "sensory deprivation"—with so little to see, hear, and feel, the psychologists theorized, the astronaut might go insane. The McGill University experiments were designed to investigate this thesis, and even more extreme cases of sensory deprivation.

The perfection of brain circuit mapping had suggested the obverse experiment to the scientists of Project Round-Square. If sensory deprivation could debase a man, weaken him, and drive him out of his mind—why not attempt "sensory enhancement." Enriching a man's senses by requiring his recticular formations to accept data detected by machines—the total memory storage of a computer, the complete electromagnetic spectrum, the sorting out of patterns and wave forms which was possible to oscilloscopes.

Volunteers weren't too hard to find—the McGill men had found volunteers, promising not much more than money and a chance at madness.

A man's concepts of the world vary according to the data he receives. For thousands of years, men had been building up systems and structures of describing the universe, without trying to improve the methods used to accept the data. Scientific instruments were not enough—could light and color have meaning, be *real,* to a blind man? The scientists of Project Round-Square hoped that the contradictions and impossibilities of reality might disappear for a man with enhanced senses. The system philosophy was illustrated in the poem, "The Blind Man and the Elephant."

"Could a dolphin discover relativity?" said Mandel, almost angrily. "Of course not—plenty of brains, just never was even able to sense much beyond the other dolphins. Likewise, there may be enormous fields of knowledge we've never noticed, because of our sensory lacks."

"More than that, it's a *positive* approach!" cried Mandel. "The first in seventy years. Before this, all of psychology was concerned with debasing man, turning him into a super rat, a little black box which was fed a stimulus and kicked back a response. Automatons!"

"What about psychoanalysis and the Freudians?"

"Bleugh!" cried Mandel, enervated himself for once. "They're the worst of all. The Id, Ego, and Super-Ego are just mental mechanisms, things beyond our control, which interact to produce behavior."

It was nice to see Mandel cheerful. There's nothing better for these neurotic types than to let them talk and talk and talk—it helps reassure them. Maybe they think no one will threaten them or kick them in the belly as long as they are blabbing. A false notion produced by too much well-written TV drama.

"But still," I said slowly, waggling a finger at him. (Or was I waggling and the finger standing still? I must drink less.)

"But, Harry," I continued. "All the colors of the rainbow won't alter this picture," and I flipped one of the White Sands Bar's napkins at him. It has a nice colored picture of a Valkyrie missile on it, from the days when White Sands was a proving ground, before the "Space Flight Bubble" burst.

"Maybe not," he mumbled. "But that's only the neurological part of the idea. We're building compulsions to succeed into it, too. The kids will *have* to work out a world without impossibilities."

"Kids?" asked Colby dimly. "Keep talking, Harry."

Mandel blinked and then continued. A lot of the rest was whined and mumbled, but I thought I got most of it.

The subjects of the brain wiring were youngsters between twelve and sixteen. The psychologists had settled on those as the optimum age limits: young enough to be typified by directness, immediacy, wholeness, spontaneity, and integral fantasy. Teenage hotshots, in other words. Old enough to want to make sense of all the data, and young enough so their world view wasn't rigid.

You can look up all the psychology words except the last. "Integral fantasy" was the most important. Studies had shown this quality is most typical of real genius, and the kids had been specially selected for it. What it means is this. Most people have fantasies, but the fantasy is "disassociated"—it is unreal to them, like sex magazines and comic books. Children, especially geniuses, have "integral fantasies"—they get wild complex ideas about the real world. Ordinary people call these "strokes of genius," if they happen to work.

"And what's more," muttered Mandel, head down on the table, "they'll have to make *sense* out of it. They've had hypnosis and drug compulsions to succeed, so their new sensory picture will have to be free of logical contradictions. It'll have to be!"

I was about to ask him how they'd be able to communicate with them after the kids had "made contact"—but before I could, Harry dropped out of the game. He lay unconscious across the table. "Kids!" he murmured, balefully.

So I was doing mathematics again, setting up Urbont matrices, the curious descendants of time-variable, multi-port switching and communications math. Far more subtle than any of these, however—the Urbont equations didn't analyze radars or satellite radio links, they symbolized the neuron patterns of the human brain.

It was tedious, subtle, absolutely-right-the-first-time work. The basic units are discrete—on-off switching conditions apply, rather than continuity. In other words, there was no margin for error.

I was getting more of those flashings of hatred and self-hatred. In my little air-conditioned cubicle in the Computer Center, I would get daymares where I would be a bug in a compartment of an ice cube tray—so cool and comfortable and . . . dead. Every so often my friends would stop by—Dick Colby, bemused and apologetic; Harry Mandel, confused and sullen.

More than once, I thought of informing the medical staff of Mandel's problems but I was afraid to. In the world of science, each man has a "paper shadow" that follows him around—dossiers, transcripts, evaluations by supervisors. Get something bad in among those papers—instability, erratic work habits, even extravagant praise, and you're in trouble. The big corporations like their scientists *a little* peculiar—just for identification. Anything serious can really ruin a man's career. I thought it might be best for Harry to take his chances—when I thought about him at all. I tried to stay away from emotional subjects.

Fortunately, about this time, the Virgin Research Corporation brought in some new entertainment, so I could relax without raising the alcohol content of my blood.

These were the Moebius movies, the new cyclic films. I'd seen the first one, *The Endless War,* previewed in New York City, a few years before. Since then, their sophistication had increased manyfold.

The basic notion was simple. The films were written in such a way that there was no beginning and no end. But this was more than a simple splicing of the two ends. Literally, it was nearly impossible to know where the story commenced. In *The Endless War* there were at least a dozen places where you could enter, stay the two hours, and leave, coming away with the impression of a complete drama. In fact, depending on where you came in, the film might have been a comedy, a tragedy, a documentary, or most anything else.

To the Nth Generation (or *Incest on It!*) was a typical improvement. It dealt with the romantic affairs of several families over (I think) three generations. After forty years, with the amatory relations of the members incredibly tangled, the snarl was twisted back on itself as the original characters were brought forth in and out of wedlock. It was pretty ghastly in a way, but also quite amusing. I hear the French are preparing a film which will do the same thing in two generations, and there's going to be a science-fiction picture that does it in one.

I had also heard rumors of the most recent development. Cyclic films had closed the old notion of time, the first "true" Moebius would eliminate time as an orienter. With pure dissonant sound, with only the most limited and ingenious movements, a complete showing would have the film run both backward and forward, right-to-right and right-to-left. Enthusiasts predicted it would make *Last Year at Marienbad* look like *Loony Tunes.*"

I got precious little pleasure out of *To the Nth Generation,* however. Mandel was moaning and groaning about his buddy-buddies and how much he hated kids through the whole thing. I didn't stay past the second time around—and these things are cumulative—ten revolutions of *The Endless War* had made me practically a pacifist.

The trouble was that half an hour into *Incest on It!* I fell off the edge of a cliff. On the screen the Most Beautiful Girl at Queens College was giving birth on the steps of the New York Public Library to the Nobel physicist who would father the owner of the biggest brothel in the Bronx, who in turn might (there were subtle hints) be the parent of the beautiful blond in labor on the dirty white steps.

I would always have teenage competitors! I would get older, and older, and older ("Never produced anything after twenty-seven!" "Burned out at twenty-five, I'd say!" "We keep him around for laughs, and to teach the remedial courses. Never did anything worthwhile after his thesis!") but they would always be coming; young, bright, arrogant, brilliant. I could barely keep from screaming and screaming and screaming. Instead, I made as tight a fist as I could, squeezing, the way I would press the foot rest in the dentist's chair, because it hurts and hurts, and you've got to do *something* when it hurts so much.

I had been wrong about Mandel, or partly. He, like I, would not be shamed. And he saw Project Round-Square as a betrayal by his friends, of the creation by his "buddy-buddies" of new competitors to torment him. I could understand this, though his woe was not mine. His complaints had stimulated me to see my own doom in the endless procreation of the film.

And now I recalled the "Kubie Report" in the *American Scientist:* "Some Unsolved Problems of the Scientific Career." The high incidence of nervous breakdown in the middle years, as the creative energy wore thin. Directness, immediacy, diversity, wholeness, spontaneity, integral fantasy! For these I had denied myself everything, sweating out my advanced degrees before age could touch me. And now I was old; I could feel myself rotting as I sat there. I could feel my brains inside my body: ropy, red, pulsing, tinged with age, hot and glowing inside a pile of gray, fatty, fibrous tissue—my unexercised body. Somehow I managed to get up and stumble out of the theater. Behind me, on the screen, someone was talking on his deathbed to his grandchildren and grandparents, who would turn out to be exactly the same people.

Part of the time I worked intensely at my tapes and card decks, not daring to pause, afraid to close my machine-language manuals. Other days, sometimes for hours, I could not work closely. I cut free of the job, drifted beyond the "grid" of the scientific attitude. What difference did it make, my mind gibed, if men landed on Mars, or discovered element 1,304? Particle, wave, wavicle, round-square—who cares? Science was just another "institution," like anything else. Would a man a thousand years from now laugh at me, the way my seminal engineer friends would chuckle at a scribe in 1000 A.D., who spent his little life endlessly recopying scrolls in a monastery?

It took all my tricks to get through the final weeks. The best was French sleep therapy, which I once read about in a book called *Force Yourself to Relax!* If your troubles are unbearable, knock yourself out until your subconscious has time to patch you up. I tried reading for a while, but I couldn't seem to understand C. P. Snow's two cultures. All I could recall was a passage about someone dying. Snow said that on the point of death, most people care not a whit about their intellectual failures or social lacks. But they cry out endlessly about their missed sensual interludes.

Richard Colby still visited me, but Mandel had stopped coming. In a moment of weakness I'd told him about his "paper shadow." He'd slammed out in a huff. The next day he came back, calm and chipper.

"You look better than I do."

"Of course. All my problems are solved."

"What about your psychology buddies?"

"Oh, I knew none of them were *really* my friends. But I'm set to take care of them, all right, all right."

"Yeah? How?"

"Well, you remember what you told me about my personal file. About how if they decide I'm unstable and borderline, it'll be very hard for me to get any sort of job, and I couldn't be a psychologist any more."

"The situation is something like that, at least with any large organization."

"But as long as I seem all right, all those rats will leave me alone, and even seem to be friendly."

"Well, you don't put it very well—"

"So it's all very simple. I've fixed them all right all right. I went over to chemical stores and got the components for a large bomb. Then I assembled it in the bottom drawer of my desk, in the middle of the Psychology Section of the project. And rigged it with a button detonator."

"Go on, go on!"

"Well, don't you see?"

"No! Go on!"

"It's perfectly plain what I—"

"Mandel, explain!"

"Well, as long as I'm feeling all right, I go on in a perfectly normal way. When I feel a little sick, I go into my office. But if, some day, I think I'm going really nuts, I think I'm really going to go crazy so it all will go down on my personal file and ruin me with the big organization—"

"Yes?"

"Why, then it's as if all my friends are suddenly about to become my enemies, to turn on me, to think I'm crazy and fire me and laugh at me and pity me behind my back." His eyes were mad, though his voice was perfectly level. "They won't be able to do that to Harry Mandel. They won't do it. I'll blow them all to bits first!"

It was enough for me, friend or no. I got word to Personnel, anonymously, and that night they called for Mandel. The poor guy hadn't thought to install one of his "White Collar Kamikazes" in his quarters, so they took him out on the G.E.M. *Topaz* that very night, under heavy sedation.

Was I in much better shape than Harry Mandel had been? A cheerful, hopelessly neurotic robot. I had come up through the sequence without much real thought about stuff like that. It was enough to do my work. Once, when I was drunk, I had the idea that if you had a perfect baby, you could set up the perfect program for his life, sports and studies and sex and social life, split-second-timed, an ideal existence. But it was already twenty years too late for that for me.

Or was that the easy way out, to flunk yourself and roll with the tide. Was I simply 'hiding out' in science because it was socially sanctioned and I had a talent for math?

But on the other side it gave me a pattern for my days, the stability I needed and craved. For this I might do work that I even despised. There must be some way to decide, to choose the optimum path, the really best way, before all your time's run through.

But how do you do it, how could I, of all people, do it?

It was my own brand of impossible.

Project Round-Square finished out fast.

Dick Colby and I sat in one of the electronics labs and watched as the countdown dropped to zero. Closed circuit television brought us a view of the MT-Section, a big room holding more than a dozen aleph-sub-sixes. Dr. Wilbur, the head of the machine-translation group, sat at the console of an aleph-sub-nine, the most advanced computer International Business Machines has ever turned out. (*Nobody* really understands it. It was designed by an aleph-sub-eight and the main purpose of the sixes was that collectively they kept the sub-nine from going crazy.)

Mathematical linguistics is the new "in" branch of math, like differential topology used to be, and category theory after that. Wilbur was playing it to the hilt, with ski boots, no tie, and a crappy old sports jacket. But he had a feel for the communications process that amounted to empathy. It was this rare talent that was needed, to help the nine through the clutches.

Above him on the lintel of the machine, was the proud motto of the National Programmer's Union, originally a remark by Queen Juliana of the Netherlands:

> *"I can't understand it, I can't even understand the people who can understand it."*

"Project Round-Square," said a disgusted technician. "Still seems crazy to me."

"Perhaps it was a bad choice of name," said Colby next to me. The Michigan electronics expert was tanned and calm, and cheerful as ever. "Did you ever read that poem, 'The Blind Men and the Elephant'? It's the same principle. With new senses or a new orientation, apparent contradictions in reality might disappear. It's what some people call 'thinking in other categories.'"

"Like the wavicle," said someone else.

"Yes," said Richard Colby, his face taunting as he smiled. "When physicists were studying certain particles, they found that in some situations they could be thought of as waves, and the equations worked out. In other sorts of reactions, you could think of them as particles, and the numbers and theories checked out under *that* hypothesis. So the physics men just shrugged and called 'em wavicles."

"But what does a wavicle look like—"

"I don't know. Nobody knows—but nobody knows what a round-square looks like either."

The lab grew quiet, as if Colby had said something profound. On the television screen the computers clicked and roared, the tape drives jerking abruptly in their vacuum columns.

"What about military applications—" I croaked at Colby. He seemed to be up to date. I hadn't paid much attention to all the interoffice junk we got on the project—including the "Virgin Tease" newsletter that told the lab assistant in Subsection Nine of Track Four of Approach Nineteen what progress had been made towards the noble goal of Project Sixty-Nine, of which he sometimes recalled he was a member.

"The guy that hired me told me this thing had military applications—I didn't know if he was kidding or I was in a nightmare."

Colby looked at me wide-eyed for a moment. "Well—I don't know. Of course, there doesn't seem to be any *direct* application. But neither did Einstein's equations, or "game theory" when it was developed. *Any* sort of insight into the world is likely to be militarily useful these days. It doesn't even have to be technical—remember the old German army staffs that were so successful. A simple thing like the chain of command could lick the best general who had to boss his own whole show."

"Just thinking is dangerous these days—often deadly."

"Maybe they ought to classify it," joked Colby.

"But then, I guess it always has been."

On the screen the computers continued to run. Coverage would be confined to the machine-translation lab; the psychologists on the staff had decided to have no reports from the real center of activity.

Outside, in a surface lab, were a dozen adolescents. They had been trained to their peak as scientists, well beyond Ph.D. level. Now they were being sensitized, exposed to the flood of phenomena that ordinary people never know about, because our wonderful minds are deaf and dumb to nearly everything in the universe. And hidden deep down below their sensitivities, there was a biting, burning, clawing, raving drive to master this new universe they would meet, and to see in it the death of the old human notion of opposites, contradictions, limitations. Nothing, their minds raved as they scanned new skies, nothing must be impossible.

The computers ran for ten hours on the screens, while Wilbur studied and fumed and paced and drank coffee. The subjects would now all have come out of anesthesia and would now be studying and observing the multitude of apparatus in the lab. Within it, I had been told, there was operating a demonstration of almost every major scientific phenomenon. The sensitized ones had been briefed to the limit, short of data which it was thought might stultify their world view. We could only wait now.

A red light glowed on the aleph-sub-nine, and data began pouring in. Wilbur threw himself into the operators' chair, and began chiding the immense computer, intuitively helping it arrange the data into some sort of language which might mean something to man. His face burned with concentration.

Near the end of the four-hour shift he looked at the machine oddly and tried another system of organization. He looked at it hard again. Then he triggered his secretary. The machine, which had responded to his winking-blinking notes, fed back the information in binary code in his earphones.

He sat back and closed his eyes for a few moments. Then he opened them and took up the microphone. "Communications established. They're intelligible," he said. And then, in a lower voice, "I've seen that structure somewhere before. . . ."

The cheers in the room blotted out anything further. Contact established, and the response was not gibberish! Well, never mind what it meant! We'd get that soon enough.

Premature congratulations? Well, perhaps. Remember the satellite shots, with a rocket roaring up on its own fire, swimming right into the calculated orbit, ejecting its satellite, and the little moon's radios bursting into life. *That* was when everyone cheered! Not six months later, when the miles of telemeter tape had been studied and restudied and been given meaning by sweat and genius. Nor did most people feel very bad when the scientists figured out that someone had forgotten to pull the safety tabs from the quick releases, so all the instruments were shielded and their data meaningless. . . .

The actual results of Project Round-Square took eleven months to evaluate and declassify. They came out, nicely distorted, in a copy of the "Virgin Tease" the V.R.C. mailed to me. . . .

Biologically, it was an unqualified success. The sensitized subjects had broken through to a whole new world of feeling. In physical terms, they were quasi-gods, for they could sense things we would never know. It was more than a widening—new colors and smells. They could sense forces and radiations and bodies all forms of which are ignored by men.

By as to the actual purpose of the project. . . .

The sensitives had been given what seemed to be an insoluble problem—eliminating contradiction in a world that required it in any rational description. The ingenuity of the human mind, the directors of Virgin Research had thought, might solve the problem, where the pure logic of machines had failed.

Well, they had solved it, in a way.

Wilbur had been right. The Hopi Indians, independently, had evolved a crude version of the sensitives' solution. The Hopi language did not allow for the complex tenses of the Indo-European tongues. All time and space to them was a single frozen matrix of events, in which the word "perhaps" had no equivalent, and the notion of "possible" and "impossible" was meaningless.

To ask, in Hopi, if it might rain tomorrow, was as meaningless as to ask if it was possible it had rained yesterday.

To the sensitives, that thing had "roundness" and this one had "squareness." Could there be such a thing as a round-square?

Perhaps, in the forever-fixed future. When it made its appearance, they would tell you. Meanwhile, it's meaningless to think about such things.

The semanticists were the only satisfied ones.

But this was only announced a year later. Not soon enough to save Harry Mandel, who was relieved of a multilinear annihilator, fifty yards from the sensitives lab. He'd slugged a guard and escaped from the *Topaz*. I understand he's in a security ward somewhere, still talking about his buddies while they watch him very carefully.

Dick Colby shrugged and grinned and had a drink or four with me at the celebration, then caught the evening ramjet back to Michigan Multi.

I collected my pay, scooter, and recommendations. Then, on the evening of the last day, I took a long look out across the desert. The sun was smoldering down into a pile of rust, the earth a great flat plum. By coincidence the girl that had assigned us our quarters was out on the desert too, but quite a ways away. I could see she was pregnant, and that was interesting to think about, and so I thought about it. I get a nice feeling when I see a woman with a child, if I have time for it—warmth, continuity for the race, the safe days of my very young childhood, though I can hardly remember any of them. There certainly were enough signs of it; she was well along.

I got a funny cold feeling. It was a tangential association, a silly one: what the boys used to call a "brassiere curve" in the rapids. You know what an ordinary bell curve looks like? Well, a lot of teachers use it for marking—most of the grades falling at C. Well, in the advanced sections some guys said the professors used a modified bell curve for figuring out grades—what they called a "brassiere curve." It looked like this:

—and jammed up inside the little hump, good and tight near the B-plus, A-minus grades, was the little gang of teenage hotshots—and me.

The cold expanded into my chest, numbing. I could look at the sunset, and the blond and her baby, and the labs and drafting boys and offices and all of it, and not feel anything at all. Everything was impersonal, like a diagram in a text.

An hour later I caught the G.E.M. *Emerald* back to my own multiversity. Classes wouldn't be starting for a while, but I'd figured out from the summer session catalogue that I could fit in a three-week intensive reading course in Chinese in the meantime. Chinese is the new "in" language for Ph.D.'s, like the way Russian used to be twenty years ago. It was pretty certain we'd be fighting them soon.

I figured I could fit in at least two "military application" courses into my fall program. Under the Rickover Plan I could get them tuition-exempt. What else could I do? I could use the money to pay for more programming. Computer operators are in short supply, and I would be sure of a fat income. I could be safe, nothing to worry about. Maybe I could get Virgin Research, or even the DOD itself, to pay for more math and science courses. I could plan it all out, interesting problems to work on the rest of my life, and get the government to pay for them. Wouldn't that be clever?

Shut up, shut up, shut up.

GREG BEAR

Tangents

The nut-brown boy stood in the California field, his Asian face shadowed by a hard hat, his short, stocky frame clothed in a T-shirt and a pair of brown shorts. He squinted across the hip-high grass at the spraddled old two-story ranch house, and then he whistled a few bars from a Haydn piano sonata. Out of the upper floor of the house came a man's high, frustrated "bloody hell!" and the sound of a fist slamming on a solid surface. Silence for a minute. Then, more softly, a woman's question: "Not going well?"

"No. I'm swimming in it, but I don't see it."

"The encryption?" the woman asked timidly.

"The tesseract. If it doesn't gel, it isn't aspic."

The boy squatted in the grass and listened.

"And?" the woman encouraged.

"Ah, Lauren, it's still cold broth."

The conversation stopped. The boy lay back in the grass, aware he was on private land. He had crept over the split-rail and brick-pylon fence from the new housing project across the road. School was out, and his mother—adoptive mother—did not like him around the house all day. Or at all.

He closed his eyes and imagined a huge piano keyboard and himself dancing on the keys, tapping out the Oriental-sounding D-minor scale, which suited his origins, he thought. He loved music.

He opened his eyes and saw the thin, graying lady in a tweed suit leaning over him, staring down with her brows knit.

"You're on private land," she said.

He scrambled up and brushed grass from his pants. "Sorry."

"I thought I saw someone out here. What's your name?"

"Pal," he replied.

"Is that a name?" she asked querulously.

"Pal Tremont. It's not my real name. I'm Korean."

"Then what's your real name?"

"My folks told me not to use it anymore. I'm adopted. Who are you?"

The gray woman looked him up and down. "My name is Lauren Davies," she said. "You live near here?"

He pointed across the fields at the close-packed tract homes.

"I sold the land for those homes ten years ago," she said. "I don't normally enjoy children trespassing."

"Sorry," Pal said.

"Have you had lunch?"

"No."

"Will a grilled cheese sandwich do?"

He squinted at her and nodded.

In the broad, red-brick and tile kitchen, sitting at an oak table with his shoulders barely rising above the top, he ate the mildly charred sandwich and watched Lauren Davies watching him.

"I'm trying to write about a child," she said. "It's difficult. I'm a spinster and I don't know children well."

"You're a writer?" he asked, taking a swallow of milk.

She sniffed. "Not that anyone would know."

"Is that your brother, upstairs?"

"No," she said. "That's Peter. We've been living together for twenty years."

"But you said you're a spinster—isn't that someone who's never married or never loved?" Pal asked.

"Never married. And never you mind. Peter's relationship to me is none of your concern." She put together a tray with a bowl of soup and a tuna-salad sandwich. "His lunch," she said. Without being asked, Pal trailed up the stairs after her.

"This is where Peter works," Lauren explained. Pal stood in the doorway, eyes wide. The room was filled with electronics gear, computer terminals, and industrial-gray shelving with odd cardboard sculptures sharing each level, along with books and circuit boards. She put the lunch tray on top of a cart, resting precariously on a box of floppy disks.

"Still having trouble?" she asked a thin man with his back turned toward them.

The man turned around on his swivel chair, glanced briefly at Pal, then at the lunch, and shook his head. The hair on top of his head was a rich, glossy black; on the close-cut sides, the color changed abruptly to a bright, fake-looking white. He had a small, thin nose and large green eyes. On the desk before him was a computer monitor. "We haven't been introduced," he said, pointing to Pal.

"This is Pal Tremont, a neighborhood visitor. Pal, this is Peter Tuthy. Pal's going to help me with that character we discussed."

Pal looked at the monitor curiously. Red and green lines went through some incomprehensible transformation on the screen, then repeated.

"What's a tesseract?" Pal asked, remembering the words he had heard through the window as he stood in the field.

"It's a four-dimensional analog of a cube. I'm trying to find a way to teach myself to see it in my mind's eye," Tuthy said. "Have you ever tried that?"

"No," Pal admitted.

"Here, Tuthy said, handing him the spectacles. "As in the movies."

Pal donned the spectacles and stared at the screen. "So?" he said. "It folds and unfolds. It's pretty—it sticks out at you, and then it goes away." He looked around the workshop. "Oh, wow!" In the east corner of the room a framework of aluminum pipes—rather like a plumber's dream of an easel—supported a long, disembodied piano keyboard mounted in a slim, black case. The boy ran to the keyboard. "A Tronclavier! With all the switches! My mother had me take piano lessons, but I'd rather learn on this. Can you play it?"

"I toy with it," Tuthy said, exasperated. "I toy with all sorts of electronic things. But what did you see on the screen?" He glanced up at Lauren, blinking. "I'll eat the food, I'll eat it. Now please don't bother us."

"He's supposed to be helping *me*," Lauren complained.

Peter smiled at her. "Yes, of course. I'll send him downstairs in a little while."

When Pal descended an hour later, he came into the kitchen to thank Lauren for lunch. "Peter's a real flake. He's trying to see certain directions."

"I know." Lauren said, sighing.

"I'm going home now." Pal said. "I'll be back, though . . . if it's all right with you. Peter invited me."

"I'm sure that it will be fine," Lauren replied dubiously.

"He's going to let me learn the Tronclavier." With that, Pal smiled radiantly and exited through the kitchen door.

When she retrieved the tray, she found Peter leaning back in his chair, eyes closed. The figures on the screen patiently folded and unfolded, cubes continuously passing through one another.

"What about Hockrum's work?" she asked.

"I'm on it," Peter replied, eyes still closed.

Lauren called Pal's adoptive mother on the second day to apprise them of their son's location, and the woman assured her it was quite all right. "Sometimes he's a little pest. Send him home if he causes trouble—but not right away! Give me a rest," she said, then laughed nervously.

Lauren drew her lips together tightly, thanked her and hung up.

Peter and the boy had come downstairs to sit in the kitchen, filling up paper with line drawings. "Peter's teaching me how to use his program," Pal said.

"Did you know," Tuthy said, assuming his highest Cambridge professorial tone, "that a cube, intersecting a flat plane, can be cut through a number of geometrically different cross sections?"

Pal squinted at the sketch Tuthy had made. "Sure," he said.

"If shoved through the plane, the cube can appear, to a two-dimensional creature living on the plane—let's call him a Flatlander—to be either a triangle, a rectangle, a trapezoid, a rhombus, or a square. If the two-dimensional being observes the cube being pushed through all the way, what he sees is one or more of these objects growing larger, changing shape suddenly, shrinking, and disappearing."

"Sure," Pal said, tapping his sneakered toe. "It's easy. Like in that book you showed me."

"And a sphere pushed through a plane would appear to the hapless Flatlander first as an *invisible* point (the two-dimensional surface touching the sphere, tangential), then as a circle. The circle would grow in size, then shrink back to a point and disappear again." He sketched the stick figures, looking in awe at the intrusion.

"Got it," Pal said. "Can I play with the Tronclavier now?"

"In a moment. Be patient. So what would a tesseract look like, coming into our three-dimensional space? Remember the program, now—the pictures on the monitor."

Pal looked up at the ceiling. "I don't know," he said, seeming bored.

"Try to think," Tuthy urged him.

"It would . . ." Pal held his hands out to shape an angular object. "It would look like one of those Egyptian things, but with three sides . . . or like a box. It would look like a weird-shaped box, too, not square."

"And if we turned the tesseract around?"

The doorbell rang. Pal jumped off the kitchen chair. "Is that my mom?"

"I don't think so," Lauren said. "More likely it's Hockrum." She went to the front door to answer. She returned with a small, pale man behind her. Tuthy stood and shook the man's hand. "Pal Tremont, this is Irving Hockrum," he introduced, waving his hand between them. Hockrum glanced at Pal and blinked a long, not-very-mammalian blink.

"How's the work coming?" he asked Tuthy.

"It's finished," Tuthy said. "It's upstairs. Looks like your savants are barking up the wrong logic tree." He retrieved a folder of papers and printouts and handed them to Hockrum.

Hockrum leafed through the printouts.

"I can't say this makes me happy," he said. "Still, I can't find fault. Looks like the work is up to your usual brilliant standards. I just wish you'd had it to us sooner. It would have saved me some grief—and the company quite a bit of money."

"Sorry," Tuthy said nonchalantly.

"Now I have an important bit of work for you. . . ." And Hockrum outlined another problem. Tuthy thought it over for several minutes and shook his head.

"Most difficult, Irving. Pioneering work there. It would take at least a month to see if it's even feasible."

"That's all I need to know for now—whether it's feasible. A lot's riding on this, Peter." Hockrum clasped his hands together in front of him, looking even more pale and worn than when he had entered the kitchen. "You'll let me know soon?"

"I'll get right on it," Tuthy said.

"Protégé?" he asked, pointing to Pal. There was a speculative expression on his face, not quite a leer.

"No, a friend. He's interested in music," Tuthy said. "Damned good at Mozart, in fact."

"I help with his tesseracts," Pal asserted.

"Congratulations," Hockrum said. "I hope you don't interrupt Peter's work. Peter's work is important."

Pal shook his head solemnly. "Good," Hockrum said, and then left the house to take the negative results back to his company.

Tuthy returned to his office, Pal in train. Lauren tried to work in the kitchen, sitting with fountain pen and pad of paper, but the words wouldn't come. Hockrum always worried her. She climbed the stairs and stood in the doorway of the office. She often did that; her presence did not disturb Tuthy, who could work under all sorts of conditions.

"Who was that man?" Pal was asking Tuthy.

"I work for him." Tuthy said. "He's employed by a very big electronics firm. He loans me most of the equipment I use here—the computers, the high-resolution monitors. He brings me problems and then takes my solutions back to his bosses and claims he did the work."

"That sounds stupid," Pal said. "What kind of problems?"

"Codes, encryptions. Computer security. That was my expertise, once."

"You mean, like fencerail, that sort of thing?" Pal asked, face brightening. "We learned some of that in school."

"Much more complicated, I'm afraid," Tuthy said, grinning. "Did you ever hear of the German 'Enigma,' or the 'Ultra' project?"

Pal shook his head.

"I thought not. Don't worry about it. Let's try another figure on the screen now." He called up another routine on the four-space program and sat Pal before the screen. "So what would a hypersphere look like if it intruded into our space?"

Pal thought a moment. "Kind of weird."

"Not really. You've been watching the visualizations."

"Oh, in *our* space. That's easy. It just looks like a balloon, blowing up from nothing and then shrinking again. It's harder to see what a hypersphere looks like when it's real. Reft of us, I mean."

"Reft?" Tuthy said.

"Sure. Reft and light. Dup and owwen. Whatever the directions are called."

Tuthy stared at the boy. Neither of them had noticed Lauren in the doorway. "The proper terms are *ana* and *kata,*" Tuthy said. "What does it look like?"

Pal gestured, making two wide swings with his arms. "It's like a ball, and it's like a horseshoe, depending on how you look at it. Like a balloon stung by bees, I guess, but it's smooth all over, not lumpy."

Tuthy continued to stare, then asked quietly, "You actually see it?"

"Sure," Pal said. "Isn't that what your program is supposed to do—make you see things like that?"

Tuthy nodded, flabbergasted.

"Can I play the Tronclavier now?"

Lauren backed out of the doorway. She felt she had eavesdropped on something momentous but beyond her. Tuthy came downstairs an hour later, leaving Pal to pick out Telemann on the keyboard. He sat at the kitchen table with her. "The program works," he said. "It doesn't work for me, but it works for him. He's a bloody natural." Tuthy seldom used such language. He was clearly awed. "I've just been showing him reverse-shadow figures. There's a way to have at least a sensation of seeing something rotated through the fourth dimension. Those hollow masks they use at Disneyland . . . seem to reverse in and out, depending on the lighting? Crater pictures from the moon—resemble hills instead of holes? That's what Pal calls the reversed images—hills and holes."

"And what's special about them?"

"Well, if you go along with the game and make the hollow faces seem to reverse and poke out at you, that is similar to rotating them in the fourth dimension. The features seem to reverse left and right—right eye becomes left eye, and so on. He caught on right away, and then he went off and played Haydn. He's gone through all my sheet music. The kid's a genius."

"Musical, you mean?"

He glanced directly at her and frowned. "Yes, I suppose he's remarkable at that, too. But spatial relations—coordinates and motion in a higher dimension. . . . Did you know that if you take a three-dimensional object and rotate it in the fourth dimension, it will come back with left-right reversed? There is no fixed left-right in the fourth dimension. So if I were to take my hand—" He held up his right hand, "and lift it *dup*—or drop it *owwen,* it would come back like this?" He held his left hand over his right, balled the right up into a fist, and snuck it away behind his back.

"I didn't know that," Lauren said. "What are *dup* and *owwen*?"

"That's what Pal calls movement along the fourth dimension. *Ana* and *kata* to purists. Like up and down to a Flatlander, who only comprehends left and right, back and forth."

She thought about the hands for a moment. "I still can't see it," she said.

"Neither can I," Tuthy admitted. "Our circuits are just too hardwired, I suppose."

Pal had switched the Tronclavier to a cathedral organ and wah-guitar combination and was playing variations on Pergolesi.

"Are you going to keep working for Hockrum?" Lauren asked. Tuthy didn't seem to hear her.

"It's remarkable," he murmured. "The boy just walked in here. You brought him in by accident. Remarkable."

"Do you think you can show me the direction—point it out to me?" Tuthy asked the boy three days later.

"None of my muscles move that way," he replied. "I can see it, in my head, but . . ."

"What is it like, seeing it? That direction?"

Pal squinted. "It's a lot bigger. Where we live is sort of stacked up with other places. It makes me feel lonely."

"Why?"

"Because I'm stuck here. Nobody out there pays any attention to us."

Tuthy's mouth worked. "I thought you were just intuiting those directions in your head. Are you telling me you're actually *seeing* out there?"

"Yeah. There's people out there, too. Well, not people, exactly. But it isn't my eyes that see them. Eyes are like muscles—they can't point those ways. But the head—the brain, I guess—can."

"Bloody hell," Tuthy said. He blinked and recovered. "Excuse me. That's rude. Can you show me the people . . . on the screen?"

"Shadows, like we were talking about."

"Fine. Then draw the shadows for me."

Pal sat down before the terminal, fingers pausing over the keys. "I can show you, but you have to help me with something."

"Help you with what?"

"I'd like to play music for them—out there. So they'll notice us."

"The people?"

"Yeah. They look really weird. They stand on us, sort of. They have hooks in our world. But they're tall . . . high dup. They don't notice us because we're so small, compared with them."

"Lord, Pal, I haven't the slightest idea how we'd send music out to them. . . . I'm not even sure I believe they exist."

"I'm not lying," Pal said, eyes narrowing. He turned his chair to face a "mouse" perched on a black ruled pad and used it to sketch shapes on the monitor. "Remember, these are just shadows of what they look like. Next I'll draw the dup and owwen lines to connect the shadows."

The boy shaded the shapes to make them look solid, smiling at his trick but explaining it was necessary because the projection of a four-dimensional object in normal space was, of course, three dimensional.

"They look like you take the plants in a garden and give them lots of arms and fingers . . . and it's kind of like seeing things in an aquarium," Pal explained.

After a time, Tuthy suspended his disbelief and stared in openmouthed wonder at what the boy was re-creating on the monitor.

"I think you're wasting your time, that's what I think," Hockrum said. "I needed that feasibility judgment by today." He paced around the living room before falling as heavily as his light frame permitted into a chair.

"I *have* been distracted," Tuthy admitted.

"By that boy?"

"Yes, actually. Quite a talented fellow."

"Listen, this is going to mean a lot of trouble for me. I guaranteed the judgment would be made by today. It'll make me look bad." Hockrum screwed up his face in frustration. "What in hell are you doing with that boy?"

"Teaching him, actually. Or rather, he's teaching me. Right now, we're building a four-dimensional cone, part of a speaker system. The cone is three dimensional—the material part—but the magnetic field forms a fourth-dimensional extension."

"Did you ever think how it looks, Peter?"

"It looks very strange on the monitor, I grant you—"

"I'm talking about you and the boy."

Tuthy's bright, interested expression fell slowly into long, deep-lined dismay. "I don't know what you mean."

"I know a lot about you, Peter. Where you come from, why you had to leave. . . . It just doesn't look good."

Tuthy's face flushed crimson.

"Keep him away," Hockrum advised.

Tuthy stood. "I want you out of this house," he said quietly. "Our relationship is at an end."

"I swear," Hockrum said, his voice low and calm, staring up at Tuthy from under his brows, "I'll tell the boy's parents. Do you think they'd want their kid hanging around an old—pardon the expression—queer? I'll tell them if you don't get the feasibility judgment made. I think you can do it by the end of this week—two days. Don't you?"

"No, I don't think so." Tuthy said softly. "Leave."

"I know you're here illegally. There's no record of you entering the country. With the problems you had in England, you're certainly not a desirable alien. I'll pass word to the INS. You'll be deported."

"There isn't time to do the work," Tuthy said.

"Make time. Instead of 'educating' that kid."

"Get out of here."

"Two days, Peter."

Over dinner, Tuthy explained to Lauren the exchange he had had with Hockrum. "He thinks I'm buggering Pal. Unspeakable bastard. I will never work for him again."

"I'd better talk to a lawyer, then," Lauren said. "You're sure you can't make him . . . happy, stop all this trouble?"

"I could solve his little problem for him in just a few hours. But I don't want to see him or speak to him again."

"He'll take your equipment away."

Tuthy blinked and waved one hand through the air helplessly.

"Then we'll just have to work fast, won't we? Ah, Lauren, you were a fool to bring me over here. You should have left me to rot."

"They ignored everything you did for them," Lauren said bitterly. She stared through the kitchen window at the overcast sky and woods outside. "You saved their hides during the war, and then . . . they would have shut you up in prison."

The cone lay on the table near the window, bathed in morning sun, connected to both the minicomputer and the Tronclavier. Pal arranged the score he had composed on a music stand before the synthesizer. "It's like a Bach canon," he said, "but it'll play better for them. It has a kind of counterpoint or over-rhythm that I'll play on the dup part of the speaker."

"Why are we doing this, Pal?" Tuthy asked as the boy sat down to the keyboard.

"You don't belong here, really, do you, Peter?" Pal asked. Tuthy stared at him.

"I mean, Miss Davies and you get along okay—but do you belong *here,* now?"

"What makes you think I don't belong?"

"I read some books in the school library. About the war and everything. I looked up *Enigma* and *Ultra.* I found a fellow named Peter Thornton. His picture looked like you but younger. The books made him seem like a hero."

Tuthy smiled wanly.

"But there was this note in one book. You disappeared in 1965. You were being prosecuted for something. They didn't even mention what it was you were being prosecuted for."

"I'm a homosexual," Tuthy said quietly.

"Oh. So what?"

"Lauren and I met in England, in 1964. They were going to put me in prison, Pal. We liked—love each other, so she smuggled me into the U.S. through Canada."

"But you're a homosexual. They don't like women."

"Not at all true, Pal. Lauren and I like each other very much. We could talk. She told me her dreams of being a writer, and I talked to her about mathematics and about the war. I nearly died during the war."

"Why? Were you wounded?"

"No. I worked too hard. I burned myself out and had a nervous breakdown. My lover . . . a man . . . kept me alive throughout the forties. Things were bad in England after the war. But he died in 1963. His parents came in to settle the estate, and when I contested the settlement in court, I was arrested." The lines on his face deepened, and he closed his eyes for a long moment. "I suppose I don't really belong here."

"I don't either. My folks don't care much. I don't have too many friends. I wasn't even born here, and I don't know anything about Korea."

"Play," Tuthy said, his face stony. "Let's see if they'll listen."

"Oh, they'll listen," Pal said. "It's like the way they talk to each other."

The boy ran his fingers over the keys on the Tronclavier. The cone, connected with the keyboard through the minicomputer, vibrated tinnily. For an hour, Pal paged back and forth through his composition, repeating passages and creating variations. Tuthy sat in a corner, chin in hand, listening to the mousy squeaks and squeals produced by the cone. *How much more difficult to interpret a four-dimensional sound,* he thought. *Not even visual clues.* Finally the boy stopped and wrung his hands, then stretched his arms. "They must have heard. We'll just have to wait and see." He switched the Tronclavier to automatic playback and pushed the chair away from the keyboard.

Pal stayed until dusk, then reluctantly went home. Tuthy stood in the office until midnight, listening to the tinny sounds issuing from the speaker cone. There was nothing more he could do. He ambled down the hall to his bedroom, shoulders slumped.

All night long the Tronclavier played through its preprogrammed selection of Pal's compositions. Tuthy lay in bed in his room, two doors down from Lauren's room, watching a shaft of moonlight slide across the wall. *How far would a four-dimensional being have to travel to get here?*

How far have I come to get here?

Without realizing he was asleep, he dreamed, and in his dream a wavering image of Pal appeared, gesturing with both arms as if swimming, eyes wide. *I'm okay,* the boy said without moving his lips. *Don't worry about me. . . . I'm okay. I've been back to Korea to see what it's like. It's not bad, but I like it better here. . . .*

* * *

Tuthy awoke sweating. The moon had gone down, and the room was pitch-black. In the office, the hypercone continued its distant, mouse-squeak broadcast.

Pal returned early in the morning, whistling disjointed selections from Mozart's Fourth Violin Concerto. Lauren opened the front door for him, and he ran upstairs to join Tuthy. Tuthy sat before the monitor, replaying Pal's sketch of the four-dimensional beings.

"Do you see them now?" he asked the boy.

Pal nodded. "They're closer. They're interested. Maybe we should get things ready, you know—be prepared." He squinted. "Did you ever think what a four-dimensional footprint would look like?"

Tuthy considered this for a moment. "That would be most interesting," he said. "It would be solid."

On the first floor, Lauren screamed.

Pal and Tuthy almost tumbled over each other getting downstairs. Lauren stood in the living room with her arms crossed above her bosom, one hand clamped over her mouth. The first intrusion had taken out a section of the living-room floor and the east wall.

"Really clumsy," Pal said. "One of them must have bumped it."

"The music," Tuthy said.

"What in *hell* is going on?" Lauren queried, her voice starting as a screech and ending as a roar.

"You'd better turn the music off," Tuthy elaborated.

"Why?" Pal asked, face wreathed in an excited smile.

"Maybe they don't like it."

A bright, filmy blue blob rapidly expanded to a diameter of a yard beside Tuthy, wriggled, froze, then just as rapidly vanished.

"That was like an elbow," Pal explained. "One of its arms. I think it's trying to find out where the music is coming from. I'll go upstairs."

"Turn it off!" Tuthy demanded.

"I'll play something else." The boy ran up the stairs. From the kitchen came a hideous hollow crashing, then the sound of vacuum being filled—a reverse pop, ending in a hiss—followed by a low-frequency vibration that set their teeth on edge.

The vibration caused by a four-dimensional creature *scraping* across their three-dimensional "floor." Tuthy's hands shook with excitement.

"Peter!" Lauren bellowed, all dignity gone. She unwrapped her arms and held clenched fists out as if she were ready to exercise or start boxing.

"Pal's attracted visitors," Tuthy explained.

He turned toward the stairs. The first four steps and a section of floor spun and vanished. The rush of air nearly drew him down the hole.

After regaining his balance, he kneeled to feel the precisely cut, concave edge. Below was the dark basement.

"Pal!" Tuthy called out. "Turn it *off*!"

"I'm playing something new for them," Pal shouted back. "I think they like it."

The phone rang. Tuthy was closest to the extension at the bottom of the stairs and instinctively reached out to answer. Hockrum was on the other end, screaming.

"I can't talk now—" Tuthy said. Hockrum screamed again, loud enough for Lauren to hear. Tuthy abruptly hung up. "He's been fired, I gather," he said. "He seemed angry." He stalked back three paces and turned, then ran forward and leapt the gap to the first intact step. "Can't talk." He stumbled and scrambled up the stairs, stopping on the landing. "Jesus," he said, as if something had suddenly occurred to him.

"He'll call the government," Lauren warned.

Tuthy waved that off. "I know what's happening. They're knocking chunks out of three-space, into the fourth. The fourth dimension. Like Pal says: clumsy brutes. They could kill us!"

Sitting before the Tronclavier, Pal happily played a new melody. Tuthy approached and was abruptly blocked by a thick green column, as solid as rock and with a similar texture. It vibrated and described an arc in the air. A section of the ceiling a yard wide was kicked out of three-space. Tuthy's hair lifted in the rush of wind. The column shrunk to a broomstick, and hairs sprouted all over it, writhing like snakes.

Tuthy edged around the hairy broomstick and pulled the plug on the Tronclavier. A cage of zeppelin-shaped brown sausages encircled the computer, spun, elongated to reach the ceiling, the floor, and the top of the monitor's table, and then pipped down to tiny strings and was gone.

"They can't see too clearly here," Pal said, undisturbed that his concert was over. Lauren had climbed the outside stairs and stood behind Tuthy. "Gee, I'm sorry about the damage."

In one smooth, curling motion, the Tronclavier and cone and all the wiring associated with them were peeled away as if they had been stick-on labels hastily removed from a flat surface.

"Gee," Pal said, his face suddenly registering alarm.

Then it was the boy's turn. He was removed more slowly, with greater care. The last thing to vanish was his head, which hung suspended in the air for several seconds.

"I think they liked the music," he said with a grin.

Head, grin and all, dropped away in a direction impossible for Tuthy or Lauren to follow. The room sucked air through the open door, then quietly sighed back to normal.

Lauren stood her ground for several minutes, while Tuthy wandered through what was left of the office, passing his hand through mussed hair.

"Perhaps he'll be back," Tuthy said. "I don't even know . . ." But he didn't finish. *Could a three-dimensional boy survive in a four-dimensional void, or whatever lay dup—or owwen?*

Tuthy did not object when Lauren took it upon herself to call the boy's foster parents and the police. When the police arrived, he endured the questions and accusations stoically, face immobile, and told them as much as he knew. He was not believed; nobody knew quite what to believe. Photographs were taken.

It was only a matter of time, Lauren told him, until one or the other or both of them were arrested. "Then we'll make up a story," he said. "You'll tell them it was my fault."

"I will *not,*" Lauren said. "But where *is* he?"

"I'm not positive," Tuthy said. "I think he's all right, however."

"How do you know?"

He told her about the dream.

"But that was before," she said.

"Perfectly allowable in the fourth dimension," he explained. He pointed vaguely up, then down, then shrugged.

On the last day, Tuthy spent the early morning hours bundled in an overcoat and bathrobe in the drafty office, playing his program again and again, trying to visualize *ana* and *kata*. He closed his eyes and squinted and twisted his head, intertwined his fingers and drew odd little graphs on the monitors, but it was no use. His brain was hard-wired.

Over breakfast, he reiterated to Lauren that she must put all the blame on him.

"Maybe it will all blow over," she said. "They have no case. No evidence . . . nothing."

All blow over, he mused, passing his hand over his head and grinning ironically. *How over, they'll never know.*

The doorbell rang. Tuthy went to answer it, and Lauren followed a few steps behind.

Putting it all together later, she decided that subsequent events happened in the following order:

Tuthy opened the door. Three men in gray suits, one with a briefcase, stood on the porch. "Mr. Peter Tuthy?" the tallest asked.

"Yes," Tuthy acknowledged.

A chunk of the doorframe and wall above the door vanished with a roar and a hissing pop. The three men looked up at the gap. Ignoring what was impossible, the tallest man returned his attention to Tuthy and continued, "Sir, it's our duty to take you into custody. We have information that you are in this country illegally."

"Oh?" Tuthy said.

Beside him, an irregular, filmy blue blob grew to a length of four feet and hung in the air, vibrating. The three men backed away. In the middle of the blob, Pal's head emerged, and below that, his extended arm and hand. Tuthy leaned forward to study this apparition. Pal's fingers waggled at him.

"It's fun here," Pal said. "They're friendly."

"I believe you," Tuthy said calmly.

"Mr. Tuthy," the tallest man valiantly persisted, though his voice was a squeak.

"Won't you come with me?" Pal asked.

Tuthy glanced back at Lauren. She gave him a small fraction of a nod, barely understanding what she was assenting to, and he took Pal's hand. "Tell them it was all my fault," he said again.

From his feet to his head, Peter Tuthy was peeled out of this world. Air rushed in. Half of the brass lamp to one side of the door disappeared. The INS men returned to their car with damp pants and embarrassed, deeply worried expressions, and without any further questions. They drove away, leaving Lauren to contemplate the quiet.

She did not sleep for three nights, and when she did sleep, Tuthy and Pal visited her and put the question to her.

Thank you, but I prefer it here, she replied.

It's a lot of fun, the boy insisted. *They like music.*

Lauren shook her head on the pillow and awoke. Not very far away, there was a whistling, tinny kind of sound, followed by a deep vibration. To her, it sounded like applause.

She took a deep breath and got out of bed to retrieve her notebook.

RUDY RUCKER

A New Golden Age

"It's like music," I repeated. Lady Vickers looked at me uncomprehendingly. Pale British features beneath wavy red hair, a long nose with a ripple in it.

"You can't hear mathematics," she stated. "It's just squiggles in some great dusty book." Everyone else around the small table was eating. White soup again.

I laid down my spoon. "Look at it this way. When I read a math paper it's no different than a musician reading a score. In each case the pleasure comes from the play of patterns, the harmonies and contrasts." The meat platter was going around the table now, and I speared a cutlet.

I salted it heavily and bit into the hot, greasy meat with pleasure. The food was second-rate, but it was free. The prospect of unemployment had done wonders for my appetite.

Mies van Koop joined the conversation. He had sparse curly hair and no chin. His head was like a large, thoughtful carrot with the point tucked into his tight collar. "It's a sound analogy, Fletch. But the musician can *play* his score, play it so that even a legislator . . ." he smiled and nodded donnishly to Lady Vickers. "Even a legislator can hear how beautiful Beethoven is."

"That's just what I was going to say," she added, wagging her finger at me. "I'm sure my husband has done lovely work, but the only way he knows how to show a person one of his beastly theorems is to make her swot through pages and pages of teeming little symbols."

Mies and I exchanged a look. Lord Vickers was a crank, an eccentric amateur whose work was devoid of serious mathematical interest. But it was thanks to him that Lady Vickers had bothered to come to our little conference. She was the only member of the Europarliament who had.

"Vat you think our chances are?" Rozzick asked her in the sudden silence, his mouth full of unchewed cauliflower.

"Dismal. Unless you can find some way of making your research appeal to the working man, you'll be cut out of next year's budget entirely. They need all the mathematics money for that new computer in Geneva, you know."

"We know," I said gloomily. "That's why we're holding this meeting. But it seems a little late for public relations. If only we hadn't let the government take over all the research funds."

"There's no point blaming the government," Lady Vickers said tartly. "People are simply tired of paying you mathematicians to make them feel stupid."

"Zo build the machine," Rozzick said with an emphatic bob of his bald little head.

"That's right," Mies said, "Build a machine that will play mathematics like music. Why not?"

Lady Vickers clapped her hands in delight and turned to me, "You mean you know how?"

Before I could say anything, Mies kicked me under the table. Hard. I got the message. "Well, we don't have quite all the bugs worked out . . ."

"But that's just too marvelous!" Lady Vickers gushed, pulling out a little appointment book. "Let's see . . . the vote on the math appropriation is June 4 . . . which gives us six weeks. Why don't you get your machine ready and bring it to Foxmire towards the end of May? The session is being held in London, you know, and I could bring the whole committee out to *feel* the beauty of mathematics."

I was having trouble moving my mouth. "Is planty time," Rozzick put in, his eyes twinkling.

Just then Watson caught the thread of the conversation. In the journals he was a famous mathematician . . . practically a grand old man. In conversation he was the callowest of eighteen-year-olds. "Who are you trying to kid, Fletcher?" He shook his head, and dandruff showered down on the narrow shoulders of his black suit.

"There's no way . . ." He broke off with a yelp of pain. Mies was keeping busy.

"If you're going to make that train, we'd better get going," I said to Lady Vickers with a worried glance at my watch.

"My dear me, yes," she agreed, rising with me. "We'll expect you and your machine on May 23 then?" I nodded, steering her across the room. Watson had stuck his head under the table to see what was the matter. Something was preventing him from getting it back out.

When I got back from the train station, an excited knot of people had formed around Watson, Rozzick, and Mies. Watson spotted me first, and in his shrill cracking voice called out, "Our pimp is here."

I smiled ingratiatingly and joined the group. "Watson thinks it's immoral to make mathematics a sensual experience," Mies explained. "The rest of us feel that greater exposure can only help our case."

"Where is machine?" Rozzick asked, grinning like a Tartar jack-o'-lantern.

"You know as well as I do that there is none. All I did was remark to Lady Vickers . . ."

"One must employ the direct stimulation of the brain," LaHaye put in. He was a delicate old Frenchman with a shock of luminous white hair.

I shook my head. "In the long run, maybe. But I can't quite see myself sticking needles in the committee's brainstems five weeks from now. I'm afraid the impulses are going to have to come in through normal . . ."

"Absolute Film," Rozzick said suddenly. "Hans Richter and Oskar Fischinger invented in the 1920s. Abstract patterns on screen, repeating and differentiating. Is in Warszawa archives accessible."

"Derisory!" LaHaye protested. "If we make of mathematics an exhibit, it should not be a tawdry *son et lumière*. Don't worry about needles, Dr. Fletcher. There are new field methods." He molded strange shapes in the air around his snowy head.

"He's right," Watson nodded. "The essential thing about mathematics is that it gives esthetic pleasure without coming through the senses. They've already got food and television for their eyes and ears, their gobbling mouths and grubbing hands. If we're going to give them mathematics, let's sock it to them right in the old gray matter!"

Mies had taken out his pen and a pad of paper. "What type of manifold should we use as the parameter space?"

We couldn't have done it if we'd been anywhere else but the Center. Even with their staff and laboratories it took us a month of twenty-hour days to get our first working math player built. It looked like one of those domey old hair dryers growing out of a file cabinet with dials. We called it a Moddler.

No one was very interested in being the first to get his brain mathed or modified or coddled or whatever. The others had done most of the actual work, so I had to volunteer.

Watson, LaHaye, Rozzick, and Mies were all there when I snugged the Moddler's helmet down over my ears. I squeezed the switch on and let the electrical vortex fields swirl into my head.

We'd put together two tapes, one on Book I of Euclid's *Elements,* and the other on iterated ultrapowers of measurable cardinals. The idea was that the first tape would show people how to understand things they'd vaguely heard of . . . congruent triangles, parallel lines, and the Pythagorean theorem. The second tape was supposed to show the power and beauty of flat-out pure mathematics. It was like we had two excursions: a leisurely drive around a famous ruin, and a jolting blast down a drag strip out on the edge of town.

We'd put the first tape together in a sort of patchwork fashion, using direct brain recordings as well as artificially punched-in thought patterns. Rozzick had done most of this one. It was all visualized geometry: glowing triangles, blooming circles, and the like. Sort of an internalized Absolute Film.

The final proof was lovely, but for me the most striking part was a series of food images which Rozzick had accidentally let slip into the proof that a triangle's area is one-half base times height.

"Since when are triangles covered with anchovy paste?" I asked Rozzick as Mies switched tapes.

"Is your vision clear?" LaHaye wanted to know. I looked around, blinking. Everything felt fine. I still had an afterglow of pleasure from the complex play of angles in Euclid's culminating proof that the square of the hypotenuse is equal to the sum of the squares on the two sides.

Then they switched on the second tape. Watson was the only one of us who had really mastered the Kunen paper on which this tape was based. But he'd refused to have his brain patterns taped. Instead he'd constructed the whole thing as an artificial design in our parameter space.

The tape played in my head without words or pictures. There was a measurable cardinal. Suddenly I knew its properties in the same unspoken way that I knew my own body. I did something to the cardinal and it transformed itself, changing the concepts clustered around it. This happened over and over. With a feeling of light-headedness, I felt myself moving outside of this endless self-transformation . . . comprehending it from the outside. I picked out a certain subconstellation of the whole process and swathed it in its logical hull. Suddenly I understood a theorem I had always wondered about.

When the tape ended I begged my colleagues for an hour of privacy. I had to think about iterated ultrapowers some more. I rushed to the library and got out Kunen's paper. But the lucidity was gone. I started to stumble over the notation, the subscripts and superscripts; I was stumped by the gappy proofs; I kept forgetting the definitions. Already the actual content of the main theorem eluded me. I realized then that the Moddler was a success. You could *enjoy* mathematics—even the mathematics you couldn't normally *understand.*

We all got a little drunk that night. Somewhere towards midnight I found myself walking along the edge of the woods with Mies. He was humming softly, beating time with gentle nods of his head.

We stopped while I lit my thirtieth cigarette of the day. In the match's flare I thought I caught something odd in Mies's expression. "What is it?" I asked, exhaling smoke.

"The music . . ." he began. "The music most people listen to is not good."

I didn't see what he was getting at, and started my usual defense of rock music.

"Muzak," Mies interrupted. "Isn't that what you call it . . . what they play in airports?"

"Yeah. Easy listening."

"Do you really expect that the official taste in mathematics will be any better? If everyone were to sit under the Moddler . . . what kind of mathematics would they ask for?"

I shrank from his suggestion. "Don't worry, Mies. There are objective standards of mathematical truth. No one will undermine them. We're headed for a new golden age."

LaHaye and I took the Moddler to Foxmire the next week. It was a big estate, with a hog wallow and three holes of golf between the gatehouse and the mansion. We found Lord Vickers at work on the terrace behind his house. He was thick-set and sported pop eyes set into a high forehead.

"Fletcher and LaHaye," he exclaimed. "I am honored. You arrive opportunely. Behold." He pulled a sheet of paper out of his special typewriter and handed it to me.

LaHaye was looking over my shoulder. There wasn't much to see. Vickers used his own special mathematical notation. "It would make a nice wallpaper," LaHaye chuckled, then added quickly, "Perhaps if you once explained the symbolism . . ."

Lord Vickers took the paper back with a hollow laugh. "You know very well that my symbols are all defined in my *Thematics and Metathematics* . . . a book whose acceptance you have tirelessly conspired against."

"Let's not open old wounds," I broke in. "Dr. LaHaye's remark was not seriously intended. But it illustrates a problem which every mathematician faces. The problem of communicating his work to nonspecialists, to mathematical illiterates." I went on to describe the Moddler while LaHaye left to supervise its installation in Lord Vickers' study.

"But this is fantastic," Vickers exclaimed, pacing back and forth excitedly. A large Yorkshire hog had ambled up to the edge of the terrace. I threw it an apple.

Suddenly Vickers was saying, "We must make a tape of *Thematics and Metathematics,* Dr. Fletcher." The request caught me off guard.

Vickers had printed his book privately, and had sent a copy to every mathematician in the world. I didn't know of anyone who had read it. The problem was that Vickers claimed he could do things like trisect angles with ruler and compass, give an internal consistency proof for mathematics, and so on. But we mathematicians have rigorous proofs that such things are impossible. So we knew in advance that Vickers' work contained errors, as surely as if he had claimed to have proved that he was twenty meters tall. To master his eccentric notation just to find out his specific mistakes seemed no more worthwhile than looking for the leak in a sunken ship.

But Lord Vickers had money and he had influence. I was glad LaHaye wasn't there to hear me answer, "Of course. I'd be glad to put it on tape."

And, God help me, I did. We had four days before Lady Vickers would bring the Appropriations Committee out for our demonstration. I spent all my waking time in Vickers' study, smoking his cigarettes, and punching in *Thematics and Metathematics*.

It would be nice if I could say I discovered great truths in the book, but that's not the way it was. Vickers' work was garbage, full of logical errors and needless obfuscation. I refrained from trying to fix up his mistakes, and just programmed in the patterns as they came. LaHaye flipped when he found out what I was up to. "We have prepared a feast for the mind," he complained, "And you foul the table with this, this . . ."

"Think of it as a ripe Camembert," I sighed. "And serve it last. They'll just laugh it off."

Lady Vickers was radiant when she heard I'd taped *Thematics and Metathematics*. I suggested that it was perhaps too important a work to waste on the Appropriations Committee, but she wouldn't hear of passing it up.

Counting her, there were five people on the committee. LaHaye was the one who knew how to run the Moddler, so I took a walk while he ran each of the legislators through the three tapes.

It was a hot day. I spotted some of those hogs lying on the smooth hard earth under a huge beech tree, and I wandered over to look at them. The big fellow I'd given the apple was there, and he cocked a hopeful eye at me. I spread out my empty hands, then leaned over to scratch his ears. It was peaceful with the pigs, and after a while I lay down and rested my head on my friend's stomach. Through the fresh green beech leaves I could see the taut blue sky.

Lady Vickers called me in. The committee was sitting around the study working on a couple of bottles of amontillado. Lord Vickers was at the sideboard, his back turned to me. LaHaye looked flushed and desperate.

"Well," I said.

"They didn't like the first tape . . ." LaHaye began.

"Dreary, dreary," Lady Vickers cried.

"We are not schoolchildren," another committee member put in.

I felt the floor sinking below me. "And the second tape?"

"I don't see how you can call that mathematics," Lady Vickers declaimed.

"There were no equations," someone complained.

"And it made me dizzy," another added.

"Here's to the new golden age of mathematics," Lord Vickers cried suddenly.

"To *Thematics and Metathematics,*" his wife added, lifting her glass. There was a chorus of approving remarks.

"That was the real thing."

"Plenty of logic."

"And so many symbols!"

Lord Vickers was smiling at me from across the room. "There'll be a place for you at my new institute, Fletcher."

I took a glass of sherry.

RUTH BERMAN

Professor and Colonel

"'Professor Robert Moriarty . . . had one of the great brains of the century.'"—Sherlock Holmes, "The Adventure of the Empty House"

"It is with a heavy heart that I take up my pen to write these the last words in which I shall ever record the singular gifts by which my friend Mr. Sherlock Holmes was distinguished. . . . It was my intention . . . to have said nothing of that event which has created a void in my life which the lapse of two years has done little to fill. My hand has been forced, however, by the recent letters in which Colonel James Moriarty defends the memory of his brother."—John H. Watson, M.D., "The Final Problem"

"Professor Robert Moriarty is . . . a man of great intellectual force."—William Gillette and Sir Arthur Conan Doyle, Sherlock Holmes, A Play

"The greatest schemer of all time, the organizer of every deviltry, the controlling brain of the underworld, a brain which might have made or marred the destiny of nations,—that's the man! But so aloof is he from general suspicion, so immune from criticism, so admirable in his management and self-effacement, that for those very words that you have uttered ['famous scientific criminal'] he could hale you to a court and emerge with your year's pension as a solatium for his wounded

character. Is he not the celebrated author of 'The Dynamics of an Asteroid,' a book which ascends to such rarefied heights of pure mathematics that it is said that there was no man in the scientific press capable of criticizing it? Is this a man to traduce?"—Sherlock Holmes, The Valley of Fear

In the summer of 1890 James visited Europe for the first time in many years, as escort to a diplomatic conference. A colonel on the staff looked well, the minister told him. James simplified that to RHIP and left his Indian HQ in the care of a major with alacrity. He was not allowed time to go home to England itself, but he did not object. He did not much care to see the home of his youth, built by the wealth of his father's substandard concrete, and he did not have fond memories of the series of flats they had lived in after the business failed. His parents were dead. But he wired his brothers.

The youngest could not leave his job immediately, and promised to join them later in Paris. But Robert said he could put off his students and other affairs and turned up in Versailles, sweating in the heat, fanning himself with his top hat, and shining with the delight of their reunion.

"You're well?" he said anxiously.

James laughed and said he was. Robert could never forget that he was, as it were, a replacement by birth. The older James had fallen victim to an outbreak of the typhoid and had not been expected to survive. "And you?" said James. "I thought perhaps you might be married now, since you've been doing well at your business, I think?"

"Yes, but there's too much to do. I haven't thought about much besides. . . . But what about you? There must be women in India," said Robert.

"Not many of them are English."

Robert, with his infernal logic, started to say something comical about the physiological identity of the races, but James winced, and Robert gave it up. James was almost sorry. If Robert had forced him to talk about the impossible possibility of defying the Empire's opinion and marrying a native, he could have pressed Robert to talk more about his business, whatever it was. But then James would have had to talk to his younger brother about love, a prospect that would doubtless have appalled them both.

They wandered a while in silence. Robert was not an entirely comfortable person for companionship in a formal garden. If you pointed out the beautiful sunflowers and then went on to the hollyhocks, you might go quite a distance before you realized that your companion was still transfixed by a sunflower, counting the florets in each spiral and charting the pattern of clockwise and widdershins spirals, apparently blind to the gold color and the beautiful halo made by the petals, which he called the ligulate flowers. "Yes, it's symmetrical," he agreed enthusiastically, when James tried to call his attention to the petals, or ligulate flowers, as the case might be. James gave it up and asked if Robert had gone any further with his sequel to *The Dynamics of an Asteroid.*

Robert let the sunflower go and looked sidelong at his brother. "No."

"Oh," said James, feeling startled. He knew Robert had been delighted with his success at finding an equation for the problem of several bodies. James tried to think of something sensible yet consoling to say about not continuing. "Perhaps something with more practical applications would be better? Knowing where an asteroid will be doesn't shake the Earth, I suppose?"

"That isn't the point!" said Robert. Then he shrugged. "But no one else seemed to understand, either."

"What *is* the point?" said James.

"The point is precision. If you can calculate a path, you can even tell where atoms of matter, or light, or energy will be at any point in time. Not that an atom's orbit is an asteroid's. But if we could calculate the relationship between energy and matter, we could fuel engines to—" He stopped and laughed at James's confusion. "Never mind. But it's all a question of trajectories. Asteroids move, atoms move, we move. Perhaps if we knew enough," he added thoughtfully, "we could turn fortune telling into truth and trace the entire past and future of any object sufficiently well defined." He pointed at the lines of his hand. "We all have fates of our own, and the ideal reasoner, given enough data, could read them anywhere. Although as for mine—" He broke off and looked around the garden, admiring the pyramidal and spherical trees, the paths that led the eye up and down and trapped the shape of the space around them, so that to glance in any direction was to know at once where one stood and how the surrounding landscape incorporated that point. "Do you know what they did, André Le Nôtre, and those other old seventeenth-century gardeners who planted the gardens of the Sun King?"

"Laid out Versailles," James suggested.

"They integrated space," Robert said, not bothering to acknowledge the simpler answer. "Look how those straight diagonals tell you where you are and where you can go. Look how the reflecting pools double the space, and define it, in those long straight edges. It's no wonder Descartes discovered at the same time how to define space numerically. The old medieval gardens, the kinds the monasteries cultivated, were only arithmetical, adding square to square, expanding in squares as far as their walls would let them, without any sight or sense of the whole. Here the paths give measure to the shape. What André felt is what old René learned to analyze."

"I like an English garden better," said James. "It's more natural."

"But the shapes—!" Robert stopped again. "Well, they integrated space, the seventeenth-century gardeners and mathematicians. I should like to integrate time. Who knows? Perhaps if I really knew how, I could wander in time as we wander in this garden."

"Surely not!" said James.

Robert grinned at him. "Well, probably not. But I'm not promising."

They leaned against a fountain of pink marble, for the coolness of the shade and the wind in the water. The jets lofted high over their heads like swords above the central goddess. Circles of little bronze frogs and lizards spat into the lake around her.

"But why aren't you trying?" said James at length. "If you feel that way about your work, your real work, I mean, shouldn't you be getting out of business? I can live on my pay now, you know. I've been able to for some time."

"In fact, you save everything I send you?" said Robert.

"Well . . . no. But I could, I think. I know the costs of my commission and the promotions—well, until the reforms in '71, anyhow—and supplementing my pay, it all held you back, and I regret more than I can—"

Robert brushed away this attempt to show gratitude for the long sacrifice. "Oh, it could have been worse," he said. "After all, all we got you was a place in the Indian Army. You were the one who had to go into exile. Anyhow, the business does better than you think, and I haven't been doing it just for you. Indeed, I enjoy it. The challenge is invigorating. And there would be certain practical difficulties entailed if I were to attempt to dispose of the concern. But eventually I mean to get back what I gave up for it, with interest, too."

"I don't understand."

"Research takes money. Even if it takes nothing else, just the time to sit quietly and think, it takes the money to live on while you sit and think. Most things take a deal more. Such great things are doing, and there's so little help given! That mad American, Michelson, who keeps measuring the speed of light and confounding himself and everyone else by proving there's no ether out there for it to push through—do you know how he does that?"

"With mirrors, I suppose."

"Of course," said Robert. "But then the mirrors must remain steady. Nine years back, when he tried in Potsdam, he was worrying that the footsteps on pavement a block away might be interfering with the results. It took him six years to get funding and support to try again. In '87 he set his contraption on a stone—and floated the stone in a trough of liquid mercury set on a bed of cement set on a pier set on bedrock. Digging and building cost money. So does mercury—especially if your hands happen to know that the stuff's poisonous. Or there's Mach with his air waves faster than sound, and Hertz just as interesting with his Maxwell waves. You know, the Austrian Navy turned over a cannon for the study of Mach waves, but the results aren't entirely satisfactory. Mach says what he really needs is a full laboratory set up especially for the work."

"No, I didn't know," James was starting to say, but his brother went rushing on.

"Then there's the younger generation to consider. I stopped at the Sorbonne, coming down, and met an interesting pair of Polish medical students—Poles come here to study, because the Russian government doesn't give them much chance at home."

"And they're good?"

"So-so. But the girl claims that her little sister Marie is a genius, if only she had the money to come here and go on with her education properly." He pursed his lips. "It sounds unlikely, but it could be so. Or I have a colleague teaching maths out in the wilds of New Zealand. He thinks highly of a boy named Eve, and another one, Rutherford, just behind. They're cropping up all around us. I tell you, man, the world's about to explode with new inventions and new theories. If Britain wants to keep her standing as a world power, she has to stay ahead in knowledge."

"You want to give money to foreigners to help Britain keep up?"

"I want to found a British institute of sciences with worldwide influence. I may even be able to do a little of my own research there, too, if I can get it set up in time. In any event, it will build on my earlier work, so that the work won't be forgotten and have to be done again. And my institute will see to that and much besides."

"Your institute," said James slowly. "Your name, I suppose."

"Yes. Why not?"

"You don't think it's a trifle vainglorious?"

"Oh, it wouldn't have to be called after me. The official name could be something more lofty. But it would be mine, just the same. It's a simple question of funding."

"Couldn't you do the same thing at a lower cost by donating to the Cavendish?"

"I may have to, if I don't raise enough for a competing institute. But I think the Cavendish would be the better for some competition, and I trust my own judgment of what most urgently needs funding more than I do theirs, anyway."

"And the profits of your business will do all that? Robert, I . . . forgive me, but . . . is it honest? It's not like father, is it?"

"It is not!"

"I'm sorry, but—"

"No, I understand. But you see, James, I'm in trade. I find things that people want, and I sell them to them. I traffic in a good many commodities. If it's wanted, it's worth selling. But, as any gentleman will tell you, if an officer and a gentleman needs telling, trade is ungentlemanly. I am not bound by gentlemanly ethics in it. I admit. . . ." He stopped, and eyed his brother, then he took off his hat to dabble a little water from a lizard on his brow. "I admit to sharp practices," he said at last. "But I give value for money. My customers get what they bargain for, I do assure you."

"I *am* sorry."

Robert leaned over, a little awkwardly, and hugged his brother. "Your turn," he said. "Tell me about India."

They loitered down the perspective, between the lines of flowers. At home in England, Sherlock Holmes was studying the daily newspapers and calculating patterns of crime. In Switzerland, the waters of the Reichenbach Falls plunged down the face of the mountain and into the deadly cauldron beneath, waiting for another summer, and Professor Moriarty.

ANATOLY DNIEPROV

The Maxwell Equations

1

It all began on a Saturday evening when, tired from my mathematical pursuits, I took up the local evening paper and came across this advertisement on the last page:

> **Kraftstudt & Company Ltd.**
> **accept orders**
> **from organizations and individuals**
> **for all manner of calculating,**
> **analytical, and computing work.**
> **High quality guaranteed. Apply:**
> **12 Weltstrasse**

That was just what I needed. For several weeks I had been sweating over Maxwell equations concerning the behavior of electromagnetic waves in the heterogeneous medium of a special structure. In the end I had managed by a series of approximations and simplifications to reduce the equations to a form that could be handled by an electronic computer. I already pictured myself traveling up to the capital and begging the administration of the Computer Center to do the job for me. For begging it would have to be, with the Center working full capacity on military problems and nobody there giving a damn for a provincial physicist's dabblings in the theory of radio-wave propagation.

And here was a computer center springing up in a small town like ours and advertising for custom in the local paper!

I took up the receiver to get in immediate touch with the company. It was only then I realized that apart from the address the advertisement gave no particulars. A computer center not on the telephone! It just didn't make sense. I rang up the editors.

"Sorry, but that was all we received from Kraftstudt," the secretary told me. "There was no telephone in the ad."

The Kraftstudt and Co. was not in the telephone directory either.

Burning with impatience I waited for the Monday. Whenever I looked up from those neatly penned equations concealing complicated physical processes, my thoughts would turn to Kraftstudt and Co. Men of vision, I thought. In our time and age when mankind endeavors to clothe its every idea in mathematical garbs, it would be hard to imagine a more profitable occupation.

Incidentally, who was this Kraftstudt? I had been resident in the town quite a long time but the name rang no bell. As a matter of fact, I did vaguely recollect having heard the name before. But I couldn't remember when or where, no matter how hard I jogged my memory.

Came the Monday. Pocketing the sheet of equations, I started out in search of 12 Weltstrasse. A fine drizzle forced me to take a taxi.

"It's a goodish way off," said the cabby, "beyond the river, next door to the lunatic asylum."

I nodded and off we went.

It took us about forty minutes. We passed through the town gates, went over a bridge, skirted a lake, and found ourselves in the country. Early green shoots could be seen here and there in the fields along the unmetaled road, and the car stalled between banks of mud every now and then, its back wheels skidding furiously.

Then roofs appeared, then the red brick walls of the lunatic asylum standing in a little depression and jocularly referred to in town as the Wise Men's Home.

Along the tall brick wall bristling with bits of broken glass ran a clinker lane. After a few turnings the taxi pulled up at an inconspicuous door.

"This is Number Twelve."

I was unpleasantly surprised to find that Kraftstudt Co.'s premises were in the same building as the Wise Men's Home. Surely

Herr Kraftstudt hasn't ganged up the loonies to do "all manner of mathematical work" for him, I thought—and smiled.

I pressed the doorbell. I had to wait long, the better part of five minutes. Then the door opened and a pale-faced man with thick tousled hair appeared and blinked in the daylight.

"Yes, sir?" he asked.

"Is this Kraftstudt's mathematical company?" I asked.

"Yes."

"And you advertised in the newspaper? . . ."

"Yes."

"I have some work for you."

"Please come in."

Telling the driver to wait for me, I bent my head and slipped through the door. It closed and I was plunged in complete darkness.

"Follow me, please. Mind the steps. Now to your left. More steps. Now we go up. . . ."

Holding me by the arm and talking thus, the man dragged me along dark crooked corridors, up and down flights of stairs.

Then a dim yellowish light gleamed overhead, and we climbed a steep stone staircase and emerged into a small hall.

The young man hurried behind a partition, pulled up a window open and said:

"I'm at your service."

I had a feeling of having come to the wrong place. The semi-darkness, the underground labyrinth, this windowless hall lighted by a single naked bulb high at the ceiling, all added up to a thoroughly odd impression.

I looked around in confusion.

"I'm at your service," the young man repeated, leaning out of the window.

"Why, yes. So this is the Kraftstudt and Co. computer center?"

"Yes, it is," he cut in with a trace of impatience, "I told you that before. What is your problem?"

I produced the sheet of equations from my pocket and handed it through the window.

"This is a linear approximation of those equations in their partial derivatives," I began to explain, a little uncertainly. "I want them solved at least numerically, say, right on the border line between two media. . . . This is a dispersion equation, you see, and the velocity of radio-wave propagation here changes from point to point."

Snatching the sheet from my hand the young man said brusquely:

"It's all clear. When do you want the solution?"

"What do you mean—when?" I said, surprised. "You must tell me when you can do it."

"Will tomorrow suit you?" he asked, his deep dark eyes now full on me.

"Tomorrow?"

"Yes. About noon. . . ."

"Good Lord! What a computer you've got! Fantastic speed!"

"Tomorrow at twelve you will have your solution, then. The charge will be four hundred marks. Cash."

Without saying another word I handed him the money together with my visiting card.

On our way back to the entrance the young man asked:

"So you are Professor Rauch?"

"Yes. Why?"

"Well, we always thought you'd come to us sooner or later."

"What made you think so?"

"Who else could place orders with us in this hole?"

His answer sounded fairly convincing.

I barely had time to say good-bye to him before the door was shut on me.

All the way home I thought about that strange computer center next door to a madhouse. Where and when had I heard the name of Kraftstudt?

2

The next day I waited for the noon mail with mounting impatience. When the bell rang at half past eleven I jumped up and ran to meet the postman. To my surprise I faced a slim pale girl holding an enormous blue envelope in her hand.

"Are you Professor Rauch, please?" she asked.

"Yes."

"Here's a package for you from Kraftstudt's. Please sign here."

There was only one name—mine—on the first page of the ledger that she held out for me. I signed and offered her a coin.

"Oh, no!" She flushed, murmured good-bye, and was gone.

When I glanced at the photo copies of a closely written manuscript I couldn't believe my own eyes. From an electronic computer I had expected something entirely different: long columns of characters with the values of the argument in the first column and those of the solution in the second.

But what I held in my hand was a strict and precise solution of my equations!

I ran my eye through page after page of calculations that took my breath away with their originality and sheer beauty. Whoever had done it possessed an immense mathematical knowledge to be envied by the world's foremost mathematicians. Almost all the modern armory of mathematics had been employed: the theory of linear and nonlinear differential and integral equations, the theory of the functions of a complex alternating current, and those of groups, and of plurality, and even such apparently irrelevant systems as topology, number theory, and mathematical logic.

I nearly cried out in delight when at the end of a synthesis of countless theorems, intermediate calculations, formulae, and equations the final solution emerged—a mathematical formula taking up three whole lines.

And to add a touch of the exquisite, the unknown mathematician had given himself the trouble of resolving the long formula into a simpler one. He had found a brief and precise form containing only the more elementary algebraic and trigonometric expressions.

At the very end, on a small inset, there was a graphic representation of the solution.

I could wish for nothing better. An equation which I thought could not be solved in the final form had been solved.

When I had recovered a little from my initial surprise and admiration I went through the photo copies again. Now I noticed that he who had solved my problem had been writing in great hurry and very closely as though trying to save on every scrap of paper and every second of time. Altogether he had written twenty-eight pages, and I pictured mentally what a titanic work that had been! Try and pen a letter of twenty-eight closely written pages in one day or just copy, without following the meaning, twenty-eight pages out of a book, and you will surely find it a hellish job.

But what I had in front of me was not a letter to a friend or a chapter copied out of a book. It was the solution of a most intricate mathematical problem—done in twenty-four hours.

For several hours I studied the closely written pages, my surprise mounting with each hour.

Where had Kraftstudt found such a mathematician? On what terms? Who was he? A man of genius nobody knew? Or perhaps one of those wonders of human nature that sometimes occur on the border line between the normal and the abnormal? A rare specimen Kraftstudt had unearthed in the Wise Men's Home?

Cases have been recorded of brilliant mathematicians ending their days in a lunatic asylum. Maybe my mathematician was one of those?

These questions plagued me for the rest of that day.

But one thing was clear: the problem had been solved not by a machine, but by a man, a mathematical wizard the world knew nothing about.

The next day, a little calmer, I reread the whole solution for the sheer pleasure of it this time, just as one will listen again and again to a piece of music one loves. It was so precise, so limpid, so beautiful that I decided to repeat the experiment. I decided to give Kraftstudt and Co. one more problem to solve.

That was easy, for I was never short of challenging problems, and I chose an equation which I had always thought impossible to break down so that it could be handled by a computer, let alone be finally solved.

This equation, too, dealt with radio-wave propagation, but it was a specific and very complex case. It was an equation of the type that theoretical physicists evolve for the fun of it and soon forget all about because they are much too complex and therefore of no use to anybody.

I was met by the same young man blinking in the daylight. He gave me a reluctant smile.

"I have another problem—" I began.

Nodding briefly, he again led me all the way through the dark corridors to the bleak reception hall.

Knowing the drill now, I went up to the window and handed him my equation.

"So it's not computers that do these things here?"

"As you see," he said without looking up from my equation.

"Whoever solved my first problem is quite a gifted mathematician," I said.

The young man did not say a word, deep as he was in my equation.

"Is he the only one in your employ or have you several?" I asked.

"What has that to do with your requirements? The firm guarantees—"

He had no time to finish, for at that moment the deep silence of the place was shattered by an inhuman scream. I started and listened. The sound was coming from behind the wall beyond the partition. It was like somebody being tortured. Crumpling the sheets with my problem, the young man, throwing a side glance and seizing me by the hand, dragged me to the exit.

"What was that?" I asked, panting.

"You'll have the solution the day after tomorrow, at twelve. You'll pay the bearer."

With those words he left me by my taxi.

3

It is hardly necessary to say that after this event my peace of mind was completely gone. Not for one moment could I forget that terrible scream which had seemed to shake the very stone vaults sheltering Kraftstudt and Co. Besides I was still under the shock of finding such a complicated problem solved by one man in one day. And finally I was feverishly waiting for the solution of my second problem. If this one, too, was solved, then. . . .

It was with shaking hands that two days later I received a package from the Kraftstudt's girl. By its bulk I could tell that it must contain the solution to the monstrously complicated piece of mathematics. With something akin to awe I stared at the thin creature in front of me. Then I had an idea.

"Please come in, I'll get the money for you."

"No, it's all right." She seemed frightened and in a hurry. "I'll wait outside. . . ."

"Come on in, no point in freezing outside," I said and all but dragged her into the hall. "I must have a look first to see whether the work's worth paying for."

The girl backed against the door and watched me with wide-open eyes.

"It is forbidden . . ." she whispered.

"What is?"

"To enter clients' flats. . . . Those are the instructions, sir. . . ."

"Never mind the instructions. I'm the master of this house and nobody will ever know you've entered."

"Oh, sir, but they will, and then. . . ."

"What then?" I said, coming nearer.

"Oh, it's so horrible. . . ."

Her head drooped suddenly and she sobbed.

I put a hand on her shoulder but she recoiled.

"Give me the seven hundred marks at once and I will go."

I held out the money, she snatched it and was gone.

Opening the package I nearly cried out with astonishment. For several minutes I stood there staring at the sheaf of photo paper unable to believe my own eyes. The calculations were done in a *different* hand.

Another mathematical genius! And of greater caliber than the first. The equations he had solved in an analytical form on fifty-three pages were incomparably more complicated than the ones I had handed in the first time. As I peered at the integrals, sums, variations, and other symbols of the highest realms of mathematics I had a sudden feeling of having been transferred into a strange mathematical world where difficulty had no meaning. It just didn't exist.

That mathematician, it seemed, had no more difficulty in solving my problem than we have in adding or subtracting two-digit numbers.

Several times I tore myself away from the manuscript to look up a thing in a mathematical manual or reference book. I was amazed by his skill in using the most complex theorems and proofs. His mathematical logic and methods were irreproachable. I did not doubt that had the best mathematicians of all nations and ages, such as Newton, Leibnitz, Gauss, Euler, Lobachevsky, Weierstrass and Hilbert, seen the way my problem had been solved they would have been no less surprised.

When I finished reading the manuscript I fell to thinking.

Where did Kraftstudt get these mathematicians? I was convinced now he had a whole team of them, not just two or three. Surely he couldn't have founded a computer firm employing only two or three men. How had he managed it? Why was his firm next door to a lunatic asylum? Who had uttered that inhuman scream behind the wall? And why?

"Kraftstudt, Kraftstudt . . ." hammered in my brain. Where and when had I heard that name? What was behind it? I paced up and down my study, pressing my head with my hands, tasking my memory.

Then I again sat down to that genius-inspired manuscript, delighting in it, rereading it part by part, losing myself to the world in the complexities of intermediate theorems and formulas. Suddenly I jumped up because I recalled that terrible inhuman scream once more and with it came the name of Kraftstudt.

The association was not fortuitous. No, it was inevitable. The screams of a man tortured and—Kraftstudt! These naturally went together. During the Second World War a Kraftstudt served as investigator in a Nazi concentration camp at Graz. For his part in the murders and inhuman treatment he got a life sentence at the Nuremberg trials.

I remembered the man's photo in all newspapers, in the uniform of an SS Obersturmführer, in a pince-nez, with wide-open, surprised eyes in a plump good-natured face. People wouldn't believe a man with such a face could have been a sadist. Yet detailed evidence and thorough investigation left no room for doubt.

What had happened to him since the trials? Maybe he had been released like many other war criminals?

But what had mathematics to do with it all? What was the connection between a sadistic interrogator and the solutions of differential and integral equations?

At this point the chain of my reasoning snapped, for I was powerless to connect those two links. Obviously there was a link missing somewhere. Some kind of mystery.

Hard as I beat my brains, however, I could think of nothing plausible. And then that girl who said, "They will know." How scared she was!

After a few days of tormenting guesswork I finally realized that unless I cracked the mystery I would probably crack up myself.

First of all I wanted to make sure that the Kraftstudt in question was that same war criminal.

4

Finding myself at the low door of Kraftstudt and Co. for the third time, I felt that what was to happen next would influence my whole life. For no reason I could understand then or later, I paid off the taxi and rang the bell only after the cab swung round the corner.

It seemed to me that the young man with his crumpled old-mannish face had been waiting for me. Without saying a word he took me by the hand and led me through the dark subterranean maze into the reception hall where I had been on the two previous occasions.

"Well, what brings you here this time?" he asked in what seemed to me a mocking tone of voice.

"I wish to speak to Herr Kraftstudt personally," I demanded.

"Our firm is not satisfying you in some way, Professor?" he asked.

"I wish to speak to Herr Kraftstudt," I insisted, trying not to look into his prominent black eyes, which now shone with malicious mockery.

"As you wish. It's none of my business," he said after a long scrutiny. "Wait here."

Then he disappeared through one of the doors behind the glass partition.

He was gone over half an hour and I was dozing off when a rustle came to me from a corner and out of the semi-darkness stepped a white-smocked figure with a stethoscope in hand. "A doctor," flashed through my mind. "Come to examine me. Is this really necessary to see Herr Kraftstudt?"

"Follow me," the doctor said peremptorily and I followed him, having no idea what was to happen next and why I had ever started it.

Light filtered into the long corridor in which we now were through a skylight high up somewhere. The corridor ended with a tall massive door. The doctor stopped.

"Wait here. Herr Kraftstudt will see you presently."

In about five minutes he opened the door wide for me.

"Well, let's go," he said in the tone of a man who was regretting what was going to happen.

I obediently followed him. We entered a wing with large bright windows and I shut my eyes involuntarily.

I was brought out of my momentary stupor by a sharp voice: "Why don't you come up, Professor Rauch?"

I turned to my right and saw Kraftstudt in a deep wickerwork chair, the very man whom I remembered so well from the newspaper pictures.

"You wished to see me?" he asked, without greeting me or rising from his desk. "What can I do for you?"

I controlled myself with an effort and went right up to his desk.

"So you have changed your occupation?" I asked, looking hard at him. He had aged in those fifteen years and the skin on his face had gathered into large flabby folds.

"What do you mean, Professor?" he asked, looking me over carefully.

"I had thought, Herr Kraftstudt, or rather hoped that you were still. . . ."

"Ah, I see." And he guffawed.

"Times have changed, Rauch. Incidentally, it's not so much your hopes I am interested in at present, as the reasons that brought you here."

"As you can probably guess, Hérr Kraftstudt, I have a fair knowledge of mathematics, I mean modern mathematics. I thought at first you had organized an ordinary computer center equipped with electronic machinery. However, I'm now convinced that this is not the case. In your establishment it's men who solve the problems. As only men of genius would solve them. And what is most strange—with monstrous, inhuman speed. If you like, I presumed to come and meet your mathematicians, who are indeed extraordinary men."

Kraftstudt first smiled, then began to laugh quietly, then louder and louder.

"I don't see the joke, Herr Kraftstudt," I said indignantly. "My wish appears ridiculous to you, does it? But don't you realize that anybody with an interest in mathematics would have the same wish on seeing the kind of solution I got from your firm?"

"I'm laughing at something quite different, Rauch. I'm laughing at your provincial narrow-mindedness. I'm laughing at you, Professor, a man respected in the town, whose learnedness has always boggled the imagination of immature maidens and old spinsters, at the way you hopelessly lag behind the swift strides of modern science!"

I was staggered by the insolence of that ex-Nazi interrogator.

"Listen, you," I shouted. "Only fifteen years ago your specialty was applying hot irons to innocent people. What right have you to prattle about swift strides of science? Come to that, I wished to see you to find what methods you use to force the brilliant people in your power to perform work which would take men of genius several years or perhaps all their life to do. I'm very glad I have found you. I consider it my duty as a scientist and citizen to let all the people in our town know that a former Nazi hangman has chosen as his new trade to abase men of science, men whose duty has always been to work for the good of humanity."

Kraftstudt got up from his chair and, frowning, approached me.

"Listen to me, Rauch. Take my advice and do not provoke me. I knew you would come to me sooner or later. But I never imagined you would be such an idiot. Frankly speaking, I thought I would find in you an ally, so to speak, and a helper."

"What?" I exclaimed. "First you explain to me by what honest or dishonest means you are exploiting the people who bring you profit."

Before my very eyes his face shrank into a lump of dirty-yellowish skin. The pale blue eyes behind the pince-nez turned into two slits that bore into me acidly. For a fleeting moment I had a feeling of a thing being examined by a prospective buyer.

"So you want me to explain to you how honestly our firm operates? So you're not satisfied with having your idiotic sums done for you as they should be done in the twentieth century? You want to experience for yourself what it means to be solving such problems?" he hissed, his vile face a mask pulsating with rage and hate.

"I don't believe all is aboveboard here. Your reputation is proof enough. And then I overheard one of your men screaming—"

"That's enough," Kraftstudt barked. "After all, I never asked you to come. But since you are here—and in such a mood—we'll make use of you whether you like it or not."

I had been unaware that the doctor who had brought me there was standing all the time behind me. At a signal from Kraftstudt a muscular hand closed on my mouth, and a piece of cotton soaked in something pungent was thrust under my nose.

I lost consciousness.

5

I came to slowly and realized that I was lying stretched on a bed. Voices of men in a heated argument crowded in on me. For a while all I knew was that their subject was scientific. Then, as my head cleared a little, I could understand what it was about.

"I can tell you this: your Nichols is no example. The coding of stimulation is highly individual, you know. What stimulates willpower in one man might stimulate something quite different in another. For instance, an electric impulse that gives Nichols pleasure deafens me. When I get it I've a feeling two tubes have been thrust into my ears with a couple of aircraft engines revving up at the other end."

"All the same the activity rhythm of neurone groups in the brain doesn't differ much from man to man. That's what our teacher's taking advantage of really."

"With not much success, though," a tired voice said. "Nothing beyond mathematical analysis so far."

"It's all a matter of time. No shortcuts here. Nobody would introduce an electrode into your brain to examine the impulses, because that would damage the brain and consequently the impulses. Now a generator allows for a wide range of change in coded impulses. And that makes for experiments without damage to the brain."

"That's as may be," the tired voice demurred. "The cases of Gorin and Void don't bear you out. The former died within ten seconds of being put inside a frequency-modulated field. The latter screamed with pain, so the generator had to be switched off immediately. You seem to forget the principal thing about neurocybernetics, friends, and that is that the network of neurones in the human body affects immense numbers of synapses. The impulses these transmit have their own frequency. As soon as you are in resonance with this natural frequency your circuit gets tremendously excited. The doctor's probing blindfolded, so to speak. And that we are still alive is pure chance."

At that moment I opened my eyes. I was lying in a room that looked like a large hospital ward with beds lining the walls. In the middle stood a big deal table piled high with remnants of food, empty tins, cigarette stubs, and scraps of paper. The scene was lit dimly by electric light. I rose on my elbows and looked round. Immediately the conversation stopped.

"Where am I?" I whispered, looking over the faces staring at me.

A voice whispered, "The new chap's come to."

"Where am I?" I repeated, addressing them all.

"So you don't know?" asked a young man in his underwear, sitting upright in the bed to my right. "This is the firm of Kraftstudt, our creator and teacher."

"Creator and teacher?" I mumbled, rubbing my leaden forehead. "What do you mean—teacher? He's a war criminal."

"Crime is relative. It all depends on the purpose. If the end is noble, any action is good," trotted out my neighbour on the right.

This piece of vulgar Machiavellianism made me look at the man with renewed curiosity.

"Where did you pick up that bit of wisdom, young man?" I said, letting my feet down and facing him.

"Herr Kraftstudt is our creator and teacher," they suddenly began to chant in chorus.

So I have landed in the Wise Men's Home after all, I thought.

"Well, friends, things must be very bad for you to say a thing like that," I said, looking them over again.

"I bet the new boy has his maths in a frequency band between ninety and ninety-five cycles!" a stout fellow shouted, half-rising from his bed.

"And he'll squeal with pain at no more than 140 cycles in the uniformly accelerated pulse code!" bellowed another.

"And he'll be forced to sleep by receiving a series of eight pulses per second with a pause of two seconds after each series!"

"I am certain the new boy will develop ravenous hunger if stimulated at a frequency of 103 cycles with a logarithmic increase in the pulse power."

The worst I could imagine had happened. I was indeed among madmen. The strange thing, however, was that they all seemed to have the same obsession: the possible influence of some kind of codes and pulses on my sensations. They thronged round me goggle-eyed, shouting out figures, giving modulations and powers, betting on how I would act "inside the generator" and "between the walls" and what power I was likely to consume.

Knowing from books that madmen should not be contradicted, I decided not to start any arguments but to try and behave like one of them. So I spoke in as inoffensive a tone as possible to my neighbor on the right. He seemed just a bit more normal than the others.

"Would you please tell me what you're all talking about? I must admit I'm completely ignorant of the subject. All these codes, pulses, neurones, stimulations—"

The room shook to a burst of guffaw. The inmates reeled with laughter, holding their sides, rocking and doubling up. The laughter became hysterical when I rose in indignation to shout them down.

"Circuit Number Fourteen. Frequency eighty-five cycles! Stimulation of anger!" somebody shouted and their laughter crescendoed.

Then I sat on the bed and resolved to wait till they calmed down.

My neighbor on the right was the first to do so. Then he sat on my bed and fixed his eyes on mine.

"Do you mean to say you really don't know anything?"

"Word of honor, not a thing. I can't make head or tail of what you were saying."

"Word of honor?"

"Word of honor."

"All right. We'll believe you, though you're certainly a rare case. Deinis, get up and tell the new boy what we're here for."

"Yes, Deinis, get up and tell him all about it. Let him be as happy as we are."

"Happy?" I asked, surprised. "Are you happy?"

"Of course we are, of course we are," they all shouted. "Why, we know ourselves now. Man's highest bliss is to know himself."

"Didn't you know yourselves before?" I asked.

"Of course not. People don't know themselves. Only those who are familiar with neurocybernetics know themselves."

"Long live our teacher!" someone shouted.

"Long live our teacher!" they all shouted in automatic unison.

The man whom they called Deinis came up and sat down on the bed next to mine.

"What education have you?" he asked in a hollow, tired voice.

"I am a professor of physics."

"Do you know anything about neuropsychology?"

"Nothing at all."

"Cybernetics?"

"Almost nothing."

"Neurocybernetics and the general theory of biologic regulation?"

"Not the vaguest idea."

An exclamation of surprise sounded in the room.

"Not a chance," Deinis muttered. "*He* won't understand."

"Go on, please, I'll try my best to follow you."

"He'll understand all right after a dozen generator sessions or so," a voice said.

"I understood after five!" someone shouted.

"A couple of turns between the walls will be even better."

"Anyway, explain things to me, Deinis," I insisted, fighting down a terrible premonition.

"Well, do you understand what life is?"

For a long time I said nothing, staring at Deinis.

"Life is a complex natural phenomenon," I uttered at last.

There was a snigger. Then another. Then many more. The inmates of the ward were looking at me as though I had just uttered some obscene nonsense. Deinis shook his head disapprovingly.

"You're in a bad way. You've a lot to learn," he said.

"Tell me where I am wrong."

"Go on, Deinis, explain to him," they all shouted in unison.

"Very well. Listen. Life is constant circulation of coded electrochemical stimulations along the neurones of your organism."

I thought that over. Circulation of stimulations along neurones. I seemed to remember hearing something like that before.

"Well, carry on."

"All the sensations that go to make up your spiritual ego are nothing but electrochemical impulses that travel from receptors up to the brain to be processed, and then down to effectors."

"Yes, well?"

"All sensations of the outer world pass along the nerve fibers to the brain. Each sensation has its own code, frequency, and speed. And these three parameters determine its quality, intensity, and duration. Understand that?"

"Let's assume I do."

"Hence life is nothing more nor less than the passage of coded information along your nerve fibers. And thought is the circulation of frequency-modulated information through the neurone synapses in the central regions of the nervous system, that is, in the brain."

"I don't quite understand that," I confessed.

"It's like this. The brain is made up of close on ten thousand million neurones similar to electric relays. They are linked up into an elaborately interconnected system by fibers called axones. These conduct stimulation from neurone to neurone. It is this wandering of stimulation along the neurones that we call thought."

My premonition grew to fear.

"He won't understand a thing until he's been inside the generator or between the walls," shouted several voices at once.

"Well, let's assume you're right. What follows from that?" I said to Deinis.

"That life can be shaped at will. By means of pulse generators stimulating the corresponding codes in the neurone synapses. And that is of enormous practical importance."

"Meaning?" I asked softly, sensing that I was about to get an insight into Kraftstudt and Co.'s activities.

"That can be best explained by an example. Let us consider the stimulation of mathematical activity. Certain backward countries are at present building what are called electronic computers. The number of triggers, or relays, such machines have does not exceed five to ten thousand. The number of triggers in the mathematical areas of the human brain is in the order of one thousand million. Nobody will ever be able to build a machine with anywhere near that number."

"Well, what of it?"

"Here you are: mathematical problems can be solved much more efficiently and cheaply by a mechanism created by Mother Nature and lodged here," Deinis passed his hand across his forehead, "than by any expensive junk built for the job."

"But machines work quicker!" I exclaimed. "A neurone, as far as I remember, can be excited no more than two hundred times per second, whereas an electronic trigger can take millions of pulses. That is precisely why fast-working machines are more efficient!"

The ward rocked with laughter again. Deinis alone retained his poker face.

"You're wrong there. Neurones can be made to take impulses at any speed provided the exciter has a sufficiently high frequency. For example, an electrostatic generator operating in the pulsed condition. If you place a brain in the radiation field of such a generator it can be made to work to any speed."

"So that is the way Kraftstudt and Co. make their money, is it!" I said, jumping up from the bed.

"He is our teacher!" they all chanted again. "Repeat it, new boy. He is our teacher!"

"Leave him," Deinis ordered suddenly. "He will understand in time that Herr Kraftstudt is our teacher. He doesn't know anything yet. Listen to this, new boy. Every sensation has its own code, its intensity and duration. The sensation of happiness—fifty-five cycles per second with coded series of one hundred pulses each. The sensation of grief—sixty-two cycles with a pause of one-tenth of a second between pulses. The sensation of joy—forty-seven cycles with pulses increasing in intensity. The sensation of sadness—two hundred and three cycles, pain—one hundred twenty-three cycles, love—fourteen cycles, poetic mood—thirty-one, anger—eighty-five, fatigue—seventeen, sleepiness—eight, and so on. Coded pulses in these frequencies move along the neurones and thus you experience all the sensations I've mentioned. They can all be produced by a pulse generator created by our teacher. He has opened our eyes to the meaning of life."

These explanations made me giddy. I didn't know what to think. The man was either as mad as a hatter or really giving me a glimpse into mankind's future. I was still dizzy from the aftereffects of the drug I'd been given in Kraftstudt's study. A wave of weariness swept over me, I lay back and closed my eyes.

"He's under frequency seven to eight cycles! He wants to sleep!" someone shouted.

"Let him have his sleep. Tomorrow he'll start learning life. They'll take him inside the generator tomorrow."

"No, he'll have his specter recorded tomorrow. He might have abnormalities."

That was the last thing I heard. I slid into deep sleep.

6

The man I met the next day at first appeared to me quite pleasant and intelligent. When I was led into his study up a floor in the firm's main building he came forward to meet me, smiling broadly, hand stretched out in greeting.

"Ah, Professor Rauch. I'm indeed pleased to meet you."

Returning his greeting with restraint I inquired after his name.

"My name is Boltz, Hans Boltz. Our chief has given me an embarrassing commission—that of extending apologies to you in his name."

"Apologies? Is your chief really subject to pangs of conscience?"

"I don't know. I'm sure I don't know, Rauch. Anyway, he's extending his most sincere apologies to you for all that has happened. He lost his temper. He doesn't like being reminded of the past, you know."

I smiled wryly.

"Why, I did not come with any intention of raking in his past. My interest lay elsewhere. I wanted to meet those who so brilliantly solved—"

"Pray, be seated, Professor. That is exactly what I was going to speak to you about."

I settled in the proffered chair and studied the broadly smiling face behind the large desk. Boltz was a typical north-country German with an elongated face, fair hair, and large blue eyes. His fingers were playing with a cigarette case.

"I'm in charge of the maths department here," he said.

"You? Are you a mathematician?"

"Yes, in a way. At least I have a smattering of it."

"That means I can meet some of them through you?"

"You've already met all of them, Rauch," Boltz said.

I stared at him blank-eyed.

"You've spent a day and night with them."

I remembered the ward and its inmates with their nonsense about impulses and codes.

"Do you expect me to believe those crackpots are the brilliant mathematicians who solved my equations?"

Not waiting for a reply I broke into laughter.

"And yet they are, indeed. Your last problem was solved by a certain Deinis. As far as I know the same individual who last night gave you a lecture on neurocybernetics."

After a few moments' thought I said:

"In that case I don't understand anything. Perhaps you would explain it all to me?"

"With pleasure. Only after you've seen this." And Boltz offered me the morning paper.

I unfolded it slowly and suddenly jumped up. Looking at me from the first page was . . . my own face framed in black. Over it was the banner caption: "Tragic death of Dr. Rauch."

"What's the meaning of this, Boltz? What sort of farce is this?" I expostulated.

"Please calm yourself. It's all quite simple really. Last night when crossing the bridge over the river on your way home from a walk near the lake, you were attacked by two escaped lunatics from the Wise Men's Home, killed, mutilated, and thrown into the river. Early this morning a corpse was discovered at the dam. The clothes, personal belongings, and papers helped to identify the corpse as yours. The police called at the Home this morning and have pieced together a complete picture of your tragic death."

It was only then I looked at my clothes and realised that the suit I had on was not mine. I dived into my pockets; all the things I'd had on me were gone.

"But this is preposterous—"

"Yes, of course, I quite agree. But what can be done, Rauch, what can be done? Without you Kraftstudt and Co. may suffer a serious setback—go bust, if you like. I don't mind telling you that we are up to our eyebrows in orders. They're all military and extremely valuable. And that means round-the-clock computing. Since we completed the first batch of problems for the Defense Ministry business has just snowballed, you could say."

"And you want me to become another Deinis for you?"

"Oh, no, Rauch. Of course not."

"Then why that farce?"

"We need you as instructor in mathematics."

"Instructor?"

I jumped up again, staring wildly at Boltz. He lighted a cigarette for himself and nodded at my chair. I sat down, completely bewildered.

"We need new mathematicians, Professor Rauch. Either we get them or we'll very soon be on the rocks."

I stared at the man, who did not seem to me half as pleasant now as he'd done before. I seemed to discern traits of innate bestiality in him, faint, but coming to the fore now.

"Well, what if I refuse?" I asked.

"That would be just too bad. I'm afraid you'd have to join our—er—computer force then."

"Is that so bad?" I asked.

"It is," Boltz said firmly, standing up. "That would mean you'd finish your days in the Wise Men's Home."

Pacing up and down the room, Boltz began to speak in the tones of a lecturer addressing an audience:

"The computing abilities of the human brain are several hundred thousand times those of an electronic computer. A thousand million mathematical nerve cells plus the aids—memory, inhibition, logic, intuition, etc.—place the brain high above any conceivable machine. Yet the machine has one essential advantage."

"Which?" I asked, still not understanding what Boltz was driving at.

"If, say, a trigger or a group of triggers is out of order in an electronic machine, you can replace the valves, resistors, or capacitors and the machine will work again. But if a nerve cell or a group of nerve cells in the computing area is out of order, replacement, alas, is impossible. Unfortunately we are obliged to make brain triggers work at an increased tempo here. As a result, wear and tear, if I may call it so, is greatly accelerated. The living computers are soon used up and then—"

"What then?"

"Then the computer gets into the Home."

"But that's inhuman—and criminal," I said hotly.

Boltz stopped in front of me, placed a hand on my shoulder and, with a broad smile, said:

"Rauch, you've got to forget all those words and notions here. If you won't forget them yourself we'll have to erase them from your memory for you."

"You will never be able to do that!" I shouted, brushing away his hand.

"Deinis's lecture was wasted on you, I see. Pity. He spoke sense. Incidentally, d'you know what memory is?"

"What has that to do with our subject? Why the hell are you all buffooning here? Why—?"

"Memory, Professor Rauch, is prolonged stimulation in a group of neurones due to a positive reverse connection. In other words, memory is the electrochemical stimulation that circulates in a given group of nerve cells in your head. You, as a physicist interested in electromagnetic processes in complex media, must realize that by placing your head in the appropriate electromagnetic field we can stop that circulation in any group of neurones. Nothing could be simpler! We cannot only make you forget what you know, but make you recall what you have never known. However, it's not in our interests to resort to these—er—artificial means. We hope your common sense will prevail. The firm will be making over to you a sizable share of its dividends."

"For what services?"

"I've already told you—for teaching mathematics. We sign up classes of twenty to thirty people with an aptitude for maths—this country has an abundance of unemployed, fortunately. Then we teach them higher mathematics in the course of two to three months—"

"But that's impossible," I said, "absolutely impossible. In such a short time, I mean. . . ."

"It's *not* impossible, Rauch. Don't forget you'll be dealing with a very bright audience, uncommonly intelligent and possessing a wonderful memory for figures. We will see to that. That is in our power."

"Also by artificial means? By means of the pulse generator?" I asked.

Boltz nodded.

"Well, do you agree?"

I shut my eyes tightly and thought hard. So Deinis and the others in the ward were normal people and had been telling me the truth yesterday. So Kraftstudt and Co. had really developed a technique of commercializing human thought, willpower, and emotions by means of electromagnetic fields. I sensed Boltz's searching glance on me and knew I must hurry with my decision. It was devilishly hard to make. If I agreed I'd be speeding my students on their way to the Wise Men's Home. If I refused I'd do the same to myself.

"Do you agree?" Boltz repeated, touching me on the shoulder.

"No," I said, my mind made up. "No. I can be no accomplice to such abomination."

"As you wish," he said with a sigh. "I'm very sorry, though."

After a minute's silence he stood up briskly, went over to the door and, opening it, called out:

"Eider, Schrank, come in here!"

"What are you going to do to me?" I asked, also getting up.

"To begin with we'll record the pulse-code specter of your nervous system."

"Which means?"

"Which means we'll record the form, intensity, and frequency of the pulses responsible for your every emotional and intellectual state and make them into a chart."

"But I won't let you. I will protest. I—"

"Show the professor the way to the test laboratory," Boltz cut in indifferently and turned his back on me to look out of the window.

7

As I entered the test laboratory I had already formed the decision which was to play a crucial role in the events that followed. My line of reasoning was this. They are going to subject me to a test that will give Kraftstudt and his gang complete information on my inner self. They need this to know what electromagnetic influence to bring to bear on my nervous system to produce any emotion or sensation they want. If they are fully successful I'll be in their power beyond hope of escape. If they are not I'll retain a certain amount of free play. Which I might soon badly need. So the only hope for me is to try to fool those gangsters as much as possible. That I can do so to a degree I deduced from what a slave of Kraftstudt's said yesterday about pulse-code characteristics being individual, except where mathematical thought is concerned.

I was led into a large room cluttered up with bulky instrumentation, the whole looking like the control room of a power station. The middle of the laboratory was taken up by a control console with instrument panels and dials. To its left, behind a screen of wire mesh, towered a transformer, several generator lamps glowing red in white porcelain panels. Fixed to the wire mesh which served as a screen grid for the generator were a voltmeter and an ammeter. Their readings were used, apparently, to measure the generator's output. Close by the control console stood a cylindrical booth made up of two metallic parts, top and bottom.

As I was led up to the booth two men rose from behind the console. One of them was the same doctor who had taken me to Kraftstudt the day before, the other—a wizened old man whom I didn't know, with sparse hair disciplined into perfect smoothness on a yellow cranium.

"Failed to persuade him," the doctor said. "I knew as much. I could see at once that Rauch belonged to the strong type. You will come to a bad end, Rauch," he said to me.

"So will you," I said.

"That's as may be, but with you it's definite."

I shrugged.

"Will you go through it voluntarily or do you want us to force you?" he then asked, looking me over insolently.

"Voluntarily. As a physicist I'm even interested."

"Splendid. In that case remove your shoes and strip to the waist. I must examine you first and take your blood pressure."

I did as I was told. The first part of "registering the specter" looked like an ordinary medical checkup—breathe, stop breathing, and the rest of it.

When the examination was over the doctor said:

"Now step into the booth. You've got a mike there. Answer all my questions. I must warn you that one of the frequencies will make you feel an intense pain. But it will go as soon as you yell out."

In my bare feet I stepped on to the porcelain floor. An electric bulb flashed on overhead. The generator droned. It was operating in the low-frequency band. The tension of the field was obviously very high. I felt this by the way waves of warmth swelled and ebbed slowly through my body. Each electromagnetic pulse brought with it a strange tickling in the joints. Then my muscles began contracting and relaxing in time to the pulses.

Presently the frequency of the warmth waves was increased.

Here it goes, I thought. If only I can bear it.

When the frequency reached eight cycles per second I would want to sleep. If only I could fight it. If only I could fool the blackguards. The frequency was slowly increasing. In my mind I counted the number of warm tides per second. One, two, three, four, more, still more. . . . Then sleepiness was on me with overwhelming suddenness. I clamped my teeth together, willing myself into wakefulness. Sleep was pushing me under like an enormous clammy weight, bearing me down, loading my eyelids. It was a miracle I was still on my feet. I bit my tongue, hoping pain would help me throw off the nightmarish burden of sleep. At that moment, as if from afar, a voice came to me:

"Rauch, how do you feel?"

"Not bad, thank you. A bit cold," I lied. I didn't recognize my own voice and bit my lips and tongue as hard as I could.

"Don't you feel sleepy?"

"No," I said, though I thought I would drop into sleep the next moment. And then, abruptly, all sleepiness was gone. The frequency must have been increased beyond the first terminal threshold. I felt fresh and cheerful as after a good snooze. Now I must fall asleep, I thought, and, shutting my eyes, snored away. I heard the doctor say to his assistant:

"Odd. Sleep at ten cycles instead of eight and a half. Write it down, Pfaff," he told the old man. "Rauch, your sensations?"

I didn't reply, still snoring loudly, my muscles relaxed, knees stuck against the side of the booth.

"Let's go on with it," said the doctor. "Increase the frequency, Pfaff, will you."

In a second I "woke up." The frequency band through which I was now passing made me experience a whole gamut of emotions and changes of mood. I was sad, then gay, then happy, then utterly miserable.

Time I yelled out, I suddenly decided.

At the moment the generator's roar increased I yelled all I could, whereupon the doctor immediately ordered:

"Cut the tension! It's the first time I've met such a crazy type. Write down: pain at seventy-five cycles per second when normal people experience it at one hundred and thirty. Go on."

That frequency is still in store for me, I thought in dread. Will I be able to cope with it?

"Now, Pfaff, try the ninety-three on him."

When the frequency stabilized something entirely unexpected happened to me. I suddenly remembered the equations which had brought me to Kraftstudt and with perfect clarity visualized every stage of their solution. This is the frequency which stimulates mathematical thinking, I thought fleetingly.

"Rauch, name the first five members of the Bessel function of the second order," the doctor demanded.

I rattled off the answer. My head was crystal clear and my whole being was permeated with a wonderful feeling of knowing all and having it on my tongue's tip.

"Name the first ten places of pi."

I named them.

"Solve a cubic equation."

The doctor dictated one with unwieldy fractional coefficients.

In two or three seconds I had the solution ready, naming all the three roots.

"Let's go on. He's quite normal in this department."

Slowly the frequency increased and I felt maudlin. There was a lump in my throat and tears welled in my eyes. But I laughed. I roared with convulsive laughter as if being vigorously tickled. I laughed, while the tears rolled down my cheeks.

"Some idiotic idiosyncrasy again. In a class of his own, you might say. I at once knew him for a strong nervous type subject to neuroses. When will he whinge, I wonder?"

I "whinged" when weeping was farthest from my mood, when all of a sudden my heart was overflowing with buoyant happiness as a nuptial cup with good wine. I wanted to troll and laugh and dance for joy. All of them—Kraftstudt and Boltz and Deinis and the doctor—seemed to me capital fellows, the jolliest chaps I had ever met. It was then that, with great effort, I started to whimper and blow my nose loudly. Though ghastly inadequate, my weeping soon elicited the now familiar comments of the expert:

"Oh, what a type. All upside down. Nothing even remotely resembling the normal specter. This fellow will give us a lot of trouble."

How far is the one hundred and thirty? I thought, in abject terror, when the happy and carefree sensation had given way to a feeling of worry, ungrounded anxiety, the presage of impending doom. . . . I started humming a tune. I was doing it mechanically, with a great effort, while my heart pounded away in premonition of something terrible, something fatal and inevitable.

I at once knew when the frequency approached the one stimulating the sensation of pain. At first there was just a dull ache in the joints of the thumb on my right hand, then a sharp pain seared through an old wartime wound. This was followed by a terrible toothache spreading at once to all the teeth. Then a splitting headache added.

Blood pulsed painfully in my ears. Shall I be able to stand it? Shall I have enough willpower to overcome the nightmarish pain and not show it? People have been known after all to be done to death in torture chambers without groaning once. History has recorded cases of people dying on the faggots mute. . . .

The pain went on increasing. Finally it reached its peak and my whole body became one knot of gnawing, stinging, racking, throbbing, excruciating pain. I was all but unconscious and saw purple specks revolving before my eyes, but I remained silent.

"Your sensation, Rauch," the doctor's voice penetrated to me.

"A sensation of murderous rage," I muttered through clenched teeth. "If I only could lay my hands on you. . . ."

"Let's go on. He's completely abnormal. Everything's the other way round with him."

And when I was on the verge of passing out, ready to scream or groan, all pain was suddenly gone. There was sweat, clammy and cold, all over my body. My every muscle trembled.

Later some frequency made me see a blinding light which was there even when I shut my eyes, then I experienced wolfish hunger, heard a scale of deafening noises, felt cold as if taken out into the frost without a stitch on. But I persisted in giving the doctor wrong answers until he fumed with rage. I knew I still had one of the most terrible tests mentioned in the ward the day before coming to me: loss of willpower. It was will that had seen me through so far. It was this invisible inner force that had helped me fight the sensations created artificially by my tormentors. But they would get at it eventually with their hellish pulse generator. Now, would they be able to find I had lost it? I waited for that frequency in dread. And it came.

Suddenly I felt indifference. Indifference to being in the hands of the Kraftstudt gang, indifference to him and his associates, indifference to myself. My mind was a complete blank. The muscles felt flabby. All sensations were gone. It was a state of total physical and moral spinelessness. I couldn't force myself to think or make the slightest movement. I had no will of my own.

And yet, surviving in some remote corner of my consciousness, a tiny thought insisted: You must . . . you must . . . you must.

You must what? Why? Whatever for? "You must . . . you must . . . you must," kept on insisting what seemed to me a single nerve cell by some kind of miracle impervious to the all-powerful electromagnetic pulses that held sway over my nerves, bidding them to feel whatever those hangmen wanted.

Later, when I learned about the theory of the central encephalic system of brain activity, according to which all the nerve cells in the cortex are governed by a single master group of nerve cells, I realized that this supreme psychic authority was impervious even to the strongest outside physical and chemical influences. That must be what saved me then.

Suddenly the doctor ordered:

"You will collaborate with Kraftstudt."

I said:

"No."

"You will do all that you are told to do."

"No."

"Run your head against the wall."

"No."

"Let's go on. He's abnormal, Pfaff, but mind you we'll get at him yet."

I shammed loss of willpower just when a sensation of the strongest will flooded my whole being and I felt I could make myself do the impossible.

Checking on my "abnormalities" the doctor put me a few more questions.

"If the happiness of mankind depended on your life, would you give it?"

"Why should I?" I asked dully.

"Can you commit suicide?"

"Yes."

"Do you want to kill the war criminal Obersturmführer Kraftstudt?"

"What for?"

"Will you collaborate with us?"

"Yes."

"Damned if I can make anything out of him! I hope it's the first and the last time I have such a case to deal with. Loss of willpower at one seventy-five. Write that down. Let's go on with it."

And they went on for another half hour. Finally the frequency chart of my nervous system was complete. The doctor now knew all the frequencies by means of which I could be made to experience any sensation or mood. At least he thought he did. Actually the only genuine frequency was the one which stimulated my mathematical abilities. And that was just what I needed most. The point was that I had evolved a plan of blowing the criminal firm sky-high. And mathematics was to be my dynamite.

8

It is an established fact that hypnosis and suggestion work best on weak-willed individuals. That was how the Kraftstudt personnel instilled in the calculators—their wills generator treated—awed obedience and reverence toward their "teacher."

I, too, was to pass through an obedience course, but because of my "abnormal" specter this was postponed for a time. I required an individual approach.

While a working place was being specially set up for me I had comparative freedom to move about. I was allowed to go out of the ward into the corridor and glance into the classrooms where my colleagues studied or worked.

I was not allowed to join in the common prayers held between the walls of a huge aluminum condenser for half an hour every morning, during which Kraftstudt's victims paid homage to the firm's head. Devoid of will and thought, they dully repeated words read to them over a closed-circuit broadcast system.

"Joy and happiness lie in self-knowledge," announced the relayed voice.

"Joy and happiness lie in self-knowledge," the twelve men on bended knees repeated in chorus, their willpower destroyed by the alternating current field "between the walls."

"By understanding the mysteries of the circulation of impulses across neurone synapses we achieve joy and happiness."

". . . joy and happiness," repeated the chorus. "How wonderful that everything is so simple! What a delight it is to know that love, fear, pain, hatred, hunger, sorrow, joy are all nothing but movement of electrochemical impulses in our bodies!"

". . . in our bodies."

"How miserable he who does not know this great truth!"

". . . this great truth," repeated the slaves dully.

"Herr Kraftstudt, our teacher and savior, gave us this happiness."

". . . happiness."

"He gave us life."

"He gave us life."

I listened to this monstrous prayer, peeping through the glass-panel door of a classroom.

Inert and flabby, with eyes half-closed, the men repeated the nightmarish maxims in expressionless voices. The electric generator hardly ten paces away pumped submission into their minds robbed of resistance. Something inhuman, vile to the extreme, bestial, and at the same time exquisitely cruel was being done to them. Boggled for comparison at the sight of that herd of miserable creatures with no will of their own, my mind could only suggest dipsomania or drug addiction at their worst.

The thanksgiving over, the twelve passed into a spacious hall with rows of desks. Suspended over each desk was a round plate of aluminum forming part of a mammoth condenser. A second plate was apparently sunk in the floor.

This hall reminded me somehow of an open-air café with shaded tables. But the idyllic impression was swept away as soon as I looked at the men under the plates.

A sheet of paper setting out the problem awaited each one of them. At first the calculators looked at these in dumb incomprehension, still under the influence of the will-destroying frequency. Presently the frequency of ninety-three cycles was switched on and a crisp order to begin work relayed.

And all the twelve, snapping up pad and pencil, pitched into feverish scribbling. This could not be called work. It was frenzy, a kind of mathematical epilepsy. The men writhed and squirmed over their pads; their hands shuttled to and fro till they blurred; their faces turned deep purple with the strain; their eyes started out of their sockets.

This lasted for the best part of an hour. Then, when their hands started moving jerkily, heads lowered almost to the tabletops and livid veins swelled ropelike in their extended necks, the generator was switched to eight cycles. All the twelve at once dropped asleep.

Kraftstudt saw to it that his slaves got some rest!

Then it all began afresh.

One day, while watching this horrible scene of mass mathematical frenzy I saw one of the calculators break down. Suddenly he stopped writing, crazily turned to one of his furiously writing neighbors and stared at him blankly for a while as if at great pains to remember something.

Then he gave a terrible guttural cry and began tearing his clothes. He bit himself, gnawed at his fingers, tore skin off his chest, battered his head on the table. Finally he passed out and slumped down on the floor.

The rest paid not the slightest attention, their pencils still working feverishly.

I was so enraged that I started pounding on the locked door. I wanted to call out to the poor devils, tell them to have done with it, to break out and fall on their tormentors. . . .

"Don't get so worked up, Herr Rauch," I heard a calm voice beside me. It was Boltz.

"You are criminals! Look what you're doing to people. What right have you to torture them?"

He smiled his bland intellectual smile and said:

"Do you remember the myth about Achilles? The gods offered him the choice between a long but quiet life and a short but turbulent one. He chose the latter. So did those men."

"But they were not offered any choice. It's you who, aided by your pulse generator, chose to stampede them toward self-annihilation for the sake of dividends!"

Boltz laughed.

"Haven't you heard them say they are happy? And so they are. Look at the way they're working in happy abandon. Does not bliss lie in creative labor?"

"I find your arguments revolting. There is a normal tempo in human life and it is criminal to try to accelerate it."

Boltz laughed again.

"You're not exactly logical, Professor. There was a time when people traveled on foot or horseback. Nowadays they fly by jet. News used to spread from mouth to mouth, taking years to snail-pace round the globe; now radio brings events right into your home even as they happen. Present-day civilization accelerates the tempo of life artificially and you don't think it's a crime. And the host of all sorts of artificial amusements and delights, aren't they too accelerating life's tempo? So why should you consider artificial acceleration of the functions of a living organism a crime? I'm certain that these people, were they to live a natural life, would not be able to do a millionth part of what they can do now. And the meaning of life, as you know, is creative activity. You will fully appreciate that when you become one of them. Soon you will know what joy and happiness are! In fact, in two days' time. A separate room is being set up for you. You will be working there alone, because, you will excuse my saying so, you are somewhat different from normal people."

Boltz slapped me amiably on the shoulder and left me alone to ponder his inhuman philosophy.

9

In accordance with my "specter" they started my obedience training at a frequency which gave me enough willpower to achieve a feat of defiance. My first feat was easy: again I shammed loss of willpower. Kneeling down and staring ahead as vacuum-eyed as I possibly could, I repeated dully the now familiar thanksgiving balderdash. In addition a few truths about neurocybernetics were inculcated in me as a novice. They boiled down to remembering which frequencies stood for what human emotions. Out of these, two were particularly important for my plans: the one stimulating mathematical thinking and another, which, luckily, was not far from the ninety-three cycles.

My training lasted for a week, after which time I was deemed obedient enough to be put to work. The first problem I was given was analyzing the possibility of intercepting an ICBM.

It took me two hours to do. The result was not cheerful for the ministry: it couldn't be done under the conditions indicated.

The second problem, also of a military nature, was calculating a neutron beam powerful enough to set off an enemy's nuclear warheads. The answer was again cheerless. A neutron cannon as calculated would have to weigh several thousand tons.

It was indeed a delight for me to solve those problems and I must have looked as possessed as the other calculators, with the difference, however, that the generator, instead of making me an obedient tool, was infusing me with confidence and enthusiasm. A joyous feeling of being on top of the world did not leave me even during the sleep breaks. I pretended to sleep but in reality I was working out my plans of condign punishment.

When I was through with the Defense Ministry problems I began to solve in my mind (so that nobody would know) the problem most important for me, how to blow Kraftstudt and Co. sky-high.

I meant the phrase metaphorically, of course, having no dynamite and no chance of obtaining any in that prisonlike madhouse. Anyway, blasting was no part of my plan.

Since the pulse generator could stimulate any human emotion, why not try to use it, I reasoned, to rouse human dignity in its victims and make them rebel against the ex-Nazi criminals? If this were possible they would require no outside help to smash this scientifically minded gang. But was there a way to do it? Was there a way, that is, to change the frequency stimulating mathematical thinking for one that unleashed anger and hatred in man?

The generator was operated by its aged creator, Dr. Pfaff, an able engineer but apparently with a strong sadistic streak. As he obviously delighted in the perverse way his creation was used, I could not count on any help from him. Dr. Pfaff was absolutely out. The generator had to work on the frequency I required without his help or knowledge.

Now if a pulse generator is overloaded, that is, if more power is taken off than its design allows, the frequency drops. That means that by adding an extra load in the form of a resistor, a generator can be made to operate on a frequency lower than shown on the dial.

Kraftstudt and Co. exploited mathematical thinking at a frequency of ninety-three cycles per second. Anger is produced by eighty-five. That meant the frequency had to be cut down by a total of eight cycles! I started calculating an extra load to do that.

During my visit to the test laboratory I had noted the readings on the voltmeter and ammeter of the generator. Their product gave me its power. Now for the mathematical problem of an extra load. . . .

I first traced in my mind the way the gigantic condensers inside which those poor devils slaved were connected to the generator. Then, in forty minutes, I solved the pertinent Maxwell equations and did all the other, most complex calculations.

It appeared Herr Pfaff had an excess of power of only one and a half watts!

This was sufficient to calculate how a frequency of ninety-three cycles could be changed to one of eighty-five. All I had to do was to earth one of the condenser plates through a resistor of 1,350 ohms.

I nearly shouted with joy. But where could I get a length of wire of that resistance? I thought next. It had to be very exact, too, or the desired effect would not be achieved.

I feverishly cast my mind about for substitutes but could think of none. A feeling of impotence swept over me when a black plastic cup suddenly appeared in my field of vision in the act of being placed on my desk by a small trembling hand. I looked up and could barely suppress an exclamation of surprise: standing in front of me was the thin girl with frightened eyes, the one who had delivered the Kraftstudt mail to me.

"What are you doing here?" I asked under my breath.

"Working," she answered, hardly moving her lips. "So you're alive."

"Yes. I need you."

Her eyes darted about.

"Everyone in town thinks you were killed. So did I."

"You go to town?"

"Yes. Almost every day, but. . . ."

I caught her tiny hand and held it in mine.

"Tell everybody in town, especially at the university, I'm alive and kept here by force. Tell them this tonight. My friends here and myself must get help to get out."

There was terror in the girl's eyes.

"What are you saying?" she whispered. "If Herr Kraftstudt gets to know, and he can find out anything. . . ."

"How often are you interrogated?"

"Next time will be the day after tomorrow."

"You've got a whole day. Screw up your courage. Don't be afraid. Do as I tell you, please."

The girl snatched her hand away and hurried out.

There were pencils in the black cup. Ten of them altogether, of different colors for different purposes. Mechanically I took the first that came to my hand and fingered it: it was marked "2B," a very soft pencil. It had plenty of graphite, a fair conductor. Then came "3B" and "5B" pencils, then those of the "H" range, hard ones, for copying. As I fingered them my mind seethed in a turmoil of speculation. Then all of a sudden, like a flash of lightning, I remembered the specific resistances of pencil graphites: A "5H" pencil has a resistance of 2,000 ohms. The next moment I had a "5H" pencil in my hand. The problem was solved now not only mathematically but practically. There in my hand was a length of wood-enclosed graphite, with the help of which I could bring punishment to a gang of modern barbarians.

I secreted the pencil in an inner pocket as carefully as a priceless treasure. Then it occurred to me where I could get two pieces of wire, one to connect to the condenser plate over my desk, the other to the radiator in the corner, with the pencil graphite in between.

I remembered the table lamp in the ward where I lived with the other calculators. It had an electric cord which, being about five feet long, could be unwound into a forty-foot length of thin wire, which would be more than enough for the job.

I had just finished my calculations when the relayed voice announced dinner time for the calculators.

I left my solitary cell in high spirits and made for the ward. Glancing back in the corridor, I saw the doctor look with obvious displeasure at the solutions of the problems I'd been given. Apparently the fact that there was no way of intercepting an ICBM or setting off the enemy's atomic bombs by a neutron cannon was not to his liking.

He had no premonition, though, of what *could* be done with ordinary graphite from a copying pencil!

10

The table lamp I had in mind had not apparently been in use for a long time. It stood in a corner on a high stool, dusty, fly-specked, its cord coiled tight round the upright.

Early in the morning when the inmates filed out to wash, I cut off the cord with a table knife and put it in my pocket. At breakfast I pocketed a knife and when everybody went out for the prayer I locked myself in the toilet. In a matter of seconds I had skinned off the insulation sheath and exposed numerous strands of thin wire, each about five feet long. Then I split the pencil gingerly, took out the graphite core and broke off three-tenths of its length. The remaining part should have the resistance I required. I made tiny notches at either end of the graphite where I secured the wires. The resistor was ready. All that remained to be done was to connect it to the condenser plate and then earth it.

That I could do during my work.

The calculators had an eight-hour working day with ten-minute breaks after each hour. After the lunch break, at 1 P.M., the hall where they worked was as a rule visited by the Kraftstudt and Co. executives. The head of the firm used to linger in the hall for some time, obviously enjoying the sight of twelve men writhing in mathematical throes. I decided it was the best time to change the frequency.

I went to my place of work that morning with the resistor all ready in my pocket. I was walking on air. At the door I met the doctor. He had brought my problems for the day.

"Hey, sawbones, wait a minute," I called out to him.

He stopped in his tracks and looked me over, astonished.

"I'd like a word with you."

"Well, what is it?" he grunted.

"It's like this," I began. "It occurred to me while I was working yesterday to return to a conversation I had with Herr Boltz. I think I was rather rash. I wonder if you would let Herr Boltz know that I agree to teach maths to the firm's new draft."

"Good for you," he said with sincerity. "I told them that your specter being what it is you should be set up as an overseer over that mathematical manure. We badly need an efficient overseer. Your working frequencies are all different. You could just walk among them and drive the lazy or those who have slipped out of resonance."

"Why, of course, doctor. But I think I'd better stick to teaching. God witness I don't feel like bashing my head against a tabletop like I saw a chap do the other day."

"Very sensible," he agreed. "I'll be speaking to Kraftstudt. I think he will agree."

"When will I know his decision?"

"By one o'clock, I expect, when we make our round of the premises."

"Good. With your permission I'll approach you then."

He nodded and walked off. On my desk I found a sheet of paper which gave me conditions for the calculation of a new pulse generator four times more powerful than the existing one. So Kraftstudt thinks of expanding his business, I thought. Yoking to it fifty-two calculators instead of the thirteen he has now. Almost lovingly I touched the pencil graphite with bits of wire in my pocket to make sure it hadn't broken.

The conditions of the problem showed me that my calculations in connection with the existing generator were correct. My hopes for success soared. I began looking forward to lunch break. When the clock on the wall showed a quarter to one I took out my device and connected its one end to a bolt on the aluminum plate above my desk. The other end I lengthened with more pieces of wire until it was long enough to reach the radiator in the corner of the room.

The last minutes dragged painfully. At last the minute hand touched twelve. I quickly connected the wire to the radiator and strode into the corridor. Advancing towards me was Kraftstudt with Pfaff, Boltz, and the doctor in attendance. At the sight of me they broke into smiles. Boltz motioned me to join them. I did so and we all stopped at the glass door of the room where the calculators worked.

Pfaff and Kraftstudt were in front and I couldn't see what was going on inside.

"That was a wise move," Boltz whispered to me. "Herr Kraftstudt has accepted your offer. You won't regret it—"

"What's the matter?" Kraftstudt asked suddenly, turning on his retinue. Engineer Pfaff cowered, looking through the door with an odd expression on his face. My heart missed a beat.

"They're not working! They're staring about, damn 'em!" Pfaff growled.

I pressed forward and looked through the glass panel. What I saw surpassed my wildest hopes. The men who before had bent so obediently over their desks were sitting upright now, looking about them boldly, and speaking to one another in loud, resolute voices.

"It's time we put an end to it, boys. D'you realize what they're doing to us?" Deinis was saying aggressively.

"Of course we do. They've been drumming into us that we achieve happiness through their pulse generator, the bastards. I've a mind to help 'em achieve theirs!"

"What's happening there?" Kraftstudt queried threateningly.

"I've no idea," Pfaff mumbled, rolling his faded eyes. "They act as if they were normal! Why don't they go on with their work?"

Kraftstudt was livid by now.

"We won't be on time with at least five defense orders," he said through clenched teeth. "See that they immediately start working!"

Boltz snapped the lock open and our party trooped in.

"Stand up to greet your teacher and savior," Boltz said loudly.

A pregnant silence was the answer. Twelve pairs of eyes full of anger and hatred blazed in our direction. A spark was enough now to set it off. My heart sang with joy. Kraftstudt and Co. was about to bust! I stepped forward.

"What are you waiting for? The hour of delivery has come. Your happiness is in your own hands. Go on, smash this criminal gang who wanted to see you all in the madhouse!"

No sooner had I finished than the calculators rushed from their places and fell on the petrified Kraftstudt and his party. They bore Boltz and the doctor down and started throttling them. They cornered Kraftstudt, punching and kicking him. Deinis straddled over the prone Pfaff and seizing his bald head by its ears drummed it against the floor. Some tore the aluminum awnings down, others smashed windowpanes. The loudspeaker torn down by a calculator crashed to the floor, followed by the desks. The floor was strewn with sheets of calculations torn to bits.

I stood in the center of that battlefield, issuing commands:

"Now don't let Kraftstudt get away! He's a war criminal! He's the kingpin of this hell on earth where you've been worked to madness! Hold tight that scoundrel Pfaff! He designed that pulse generator! Give Boltz what he deserves! He recruited you and planned to recruit many more!"

And the men, splendid in their righteous wrath, punched, kicked, and throttled their enemies.

Though no longer under the influence of the generator they could not stop now in their noble indignation of people breaking free from thraldom. Kraftstudt and his party, torn and bleeding, were dragged into the corridor and to the exit.

I led the agitated men, hooting and jeering and cursing their former masters, through the windowless reception hall where I'd handed in my problems, through the narrow subterranean maze to the back door where we finally emerged into the open.

We were blinded momentarily by a hot spring sun and we stopped short. But not only because of the sun. In front of the door leading to Kraftstudt's apartments pressed a huge crowd of people. They had been shouting something but at the sight of us suddenly went silent. Then I heard somebody call out:

"Why, this is Professor Rauch! So he is really alive!"

Deinis and his colleagues kicked forward the battered executives of Kraftstudt and Co. One after another they struggled to their feet and glanced cowardly from us to the crowd pressing threateningly round them.

A thin pale girl broke from the crowd. So she had found courage to do what I asked her!

"That's him," she said, pointing at Kraftstudt. "And him," she added, nodding her chin at Pfaff. "They started it all. . . ."

A murmur came from the crowd. Voices were raised in anger. The people surged forward. Another moment and the criminals would have been torn limb from limb. But Deinis raised his hand.

"Friends, we're civilized people," he said. "We mustn't take justice into our own hands. The interests of humanity will be better served if we let the world know about their crimes. They must be brought to trial and we will all stand as witnesses. Within those walls heinous crimes have been committed. Taking advantage of the progress of science, those monsters were reducing men to slavery and exploiting them to the last spark of life."

"Bring the criminals to trial!" everybody shouted. "Bring 'em to trial!"

The crowd headed for town. The criminals were in a tight circle. Elsa Brinter, the thin girl, walked at my side. She clutched my hand as she spoke to me:

"I thought hard after our last conversation. Then I somehow felt strong and brave. And very angry for you and your friends and myself too."

"That's what always happens to those who hate their enemies and love their friends," I said.

Kraftstudt and his associates were handed over to the town authorities. The burgomaster made a long speech studded with biblical references. He ended by saying: "For crimes so subtle in their cruelty Herr Kraftstudt and his colleagues will be tried by the Federal Court of Justice."

Then they were taken away in police vans and have not been heard of since. Nor have there been any reports in the press. But it has been rumored that Kraftstudt and his colleagues entered government service and were entrusted with setting up a large computer center for the Defense Ministry.

I always boil with indignation now when looking through a newspaper I find on the last page this perennial advertisement:

WANTED
for work at a large computer center men aged 25-40 and having knowledge of higher mathematics. Write to Box***

MARTIN GARDNER

Left or Right?

Major Karston picked up the soggy, crumpled sheet of yellow paper and opened it carefully. His hands were shaking. Would the writing, scribbled hastily in pencil a week before, read from left to right or right to left? It was odd or even, plus or minus, a fifty-fifty chance like the toss of a coin—but the destiny of Earth was in the balance.

Our spaceship had left the earth two months ago on a crucial military mission. A surprise attack from one of the primitive cultures in the galaxy—inhabitants from the Zeta-59 planetary system—had caught the Earth in a state of woeful unpreparedness. Our stratospheric defense network had kept most of their crude missiles from getting through, but one of the rockets had slipped past, landed, and totally demolished the giant munitions plant in Alaska.

The factory had been engaged primarily in the manufacture of helixons, small but intricate spiral-shaped devices that form an essential part of our atomic shells and warheads. Its destruction had created a fatal shortage. Our only hope lay in the immediate import of a new supply.

The nearest source of helixons was a planet in the Omicron-466 system, colonized several centuries ago by our own people. It was halfway across the galaxy, but with modern methods of space travel, taking advantage so to speak of fourth-dimensional shortcuts, the trip could be made in a few weeks. Major Karston and I were chosen to command the mission.

Our cargo-transport ship, carrying a crew of seventy-five men and officers, had little difficulty evading the enemy ships that hovered outside the Earth (they were clumsy, rocket-propelled types with virtually no speed or maneuverability). The trip to the colony planet was relatively uneventful. But on our way back, while we were cruising through an uncharted portion of the galaxy, the sergeant on radar watch carelessly dozed off and the ship was struck a glancing blow by a small meteor. The blow catapulted the vessel momentarily along the fourth spatial coordinate, flipping the ship over in a series of somersaults before it dropped back into our three-dimensional continuum. Luckily no one was seriously hurt, but there had been some damage to the steering gear that made it necessary to land for repairs.

We finally located a small planet that circled about a pair of moderately sized suns, and managed a successful landing. While the crew worked feverishly on the repairs, Major Karston and I, wearing our oxygen bells and voice transmission devices, sat on the edge of a low ridge of black rock where we could watch the crew.

Suddenly Karston raised his fist, beating it violently against the glass in front of his forehead. "My God, Reilly!" he shouted. "It never occurred to me until now."

I turned and squinted at him, trying with my hands to block off the blinding light from the two suns back of his head. "What never occurred to you?"

"How many times did the ship rotate before it dropped back into our space?" he asked.

I shook my head and said, "I was too frightened to remember anything."

Major Karston's lower lip pulled on his moustache. His expression was grim. "If we turned an even number," he continued slowly, as though trying to visualize something in his mind, "we're okay. If not—" He shuddered.

"I don't understand."

"Look, my dear lieutenant," he said impatiently. "Don't be an imbecile. If you turn a two-dimensional figure on its back, it reverses the right and left sides. Right?"

"Right," I said.

"But if you turn it twice, it's the same as it was before. In other words, an odd number of turns reverses it. An even number doesn't."

"I follow," I said.

"Think back to your high-school courses in non-Euclidean manifolds," he went on. "If a three-dimensional spaceship somersaults through the fourth coordinate, turning over an odd number of times, what happens? Remember Lieutenant Semer's accident a year ago? His heart's still on the wrong side, and he hasn't even learned to read yet."

I was beginning to get the idea. "You mean that maybe our ship, including all of us, got reversed yesterday? I don't *feel* reversed."

"Of course not," he said sharply. "Your eyes, the sides of your brain, everything reverses. You may be left-handed now, but with the brain reversed you'll feel you've been that way all your life. Of course every object in the ship would have been changed too, so we haven't anything to check against."

"Until we get back, I suppose," I said. "Then if we've been reversed everything *there* will seem backward—like in a mirror."

"It isn't funny," Karston said. "Our helixons are of the plus type. That means they twist clockwise. If we return with a shipment of eight billion counterclockwise helixons, they'll be as much use to the high command as eight billion rubber thumbtacks."

Invisible fingers beat a tattoo along my backbone. "Isn't there any way we can find out?" I said.

"No. There's nothing we can do. This planet obviously hasn't enough atmosphere to support life, and even if it were inhabited by intelligent creatures it wouldn't do us any good. There's no way we could make them understand what our language means by 'left.' We haven't any charts for this region, so we can't check by star patterns. We can't learn anything from chemical tests of substances here because all the natural compounds and crystals with asymmetrical molecular structure exist in two mirror-image forms. And there aren't any physical tests because all the laws of physics are symmetrical."

"What about the left-hand rule?" I said. "The north pole of a compass always points left if you hold it under a current moving toward you. We can set up a current, and if our compasses have been reversed, the north pole will still point left. But if it seems to us to point right, we'll know we've been changed."

Karston stood up and shook his head so violently that the bell rattled. "You should review your elementary magnetism, Reilly. If our compasses are reversed, their entire atomic structure, electron spin and all, will be reversed. They'll orient themselves by a 'right-hand rule.'"

"What *can* we do?"

He spread his arms in a hopeless gesture. The sky behind him was a deep scarlet now from the setting suns, and two long shadows branched out from his feet along the dark rocky surface. "We'll have to hope for the best. Even if we are reversed—and we'll find out soon enough when we reach a charted region—there isn't any way we can somersault back again without running the risk of smashing the ship. And it's too late to go back for another shipment."

Suddenly a thought struck me. "We ejected a load of garbage and wastepaper just before the crash. If we could locate it and thaw it out, we could find those notes you made, and—"

I didn't have to finish. Karston was whacking me on the back of my space jacket and telling me I wasn't as stupid as he suspected.

We finished the repairs and shoved off the following day. After returning to the vicinity of the crash, it didn't take our radar division long to locate the batch of refuse we had left suspended in space. The material was close to absolute zero, so it was hours after we retrieved it before it had warmed enough to be examined.

Major Karston picked up one of the wadded balls of moist yellow paper. They were notes he had made for a speech in which he planned to announce the successful completion of our mission.

With trembling fingers he smoothed the paper flat and I bent over his shoulder to look. I think my heart was pounding louder than our atomic motors. Would we be able to read it? Or would it be in mirror writing, like the "Jabberwocky" in that ancient classic about the little girl who walked through the looking glass?

Odd or even, left or right, a fifty-fifty chance like the flip of a coin—but the fate of our people was hanging in the balance!

IAN WATSON

Immune Dreams

Adrian Rosen returned from Thibaud's sleep laboratory with a stronger presentiment than ever that he was about to develop cancer. He wasn't so much anxious about this as simply convinced of it as a truth—and certain, too, that in some as yet ill-defined way he was partly in control of these events about to take place inside his body . . .

"It's obsessional," Mary Strope grieved. "You're receding—from me—from reality. I wish you'd give up this line of research. This constant brooding is vile. It's ruining you."

"Maybe this recession into myself is one of the onset symptoms," Rosen meditated. "A psychological swabbing-down and anesthetizing before the experience?" He lit another of the duty-free Gitanes he'd brought back from France and considered the burning tip. The smoke had no time to form shapes, today. It was torn away too quickly by the breeze, which seemed to be smoking the cigarette on his behalf—as though weather, landscape, and his own actions concurred perfectly. The hood was down, the car open to the sky.

They sat in silence and watched the gliders being launched off the hilltop, this red-haired, angular woman (fiery hair sprouting upon a gawky frame, like a match flaring) and the short burly man with heavy black-framed sunglasses clamped protectively to his face as though he had become fragile suddenly.

The ground fell away sharply before them, to reappear as the field-checked vale far below. The winch planted a hundred yards to their right whined as it dragged a glider towards it and lofted it into the up-currents, to join two other gliders soaring a mile away among the woolpack clouds. As the club's Land-Rover drove out from the control caravan to retrieve the fallen cable, Rosen stared at the directional landing arrows cut in the thin turf, exposing the dirty white chalk—in which the ancient horse, a few miles away, was also inscribed. Beyond, a bright orange wind sock fluttered. Pointers . . .

"You don't even inhale," Mary snapped. "You could give up overnight if you were really worried."

"I know. But I won't. I'm seeing how near a certain precipice I can edge before . . . the lip gives way. It needn't be lung cancer, you know. It needn't have anything to do with cigarettes . . ."

How could he explain? His smoking was only metaphorical now. Cigarettes were a clock; a pacemaker of the impending catastrophe. In fact, he was fairly sure that it wouldn't be a smoker's cancer at all. But it sounded absurd whenever he tried to explain this.

Then there were the dreams . . .

Rosen stood before the blackboard in the seminar room of the Viral Cancer Research Unit attached to St. David's Hospital and sketched the shape of catastrophe upon it with a stick of squeaky chalk that reminded him irresistibly of school days and algebra lessons . . . The difficulty he'd had at first in comprehending *x* and *a* and *b*! His childish belief that they must equal some real number—as though it was all a secret code, and he the cryptographer! But once presented as geometry, mathematics had become crystal clear. He'd been a visualizer all along . . .

On the blackboard was the cusp catastrophe of René Thom's theorem: a cliff edge folding over, then under itself, into an overhang impossible on any world with gravity, before unfolding and flattening out again on a lower level. The shape he'd graphed was stable in two phases: its upper state and its lower state. But the sinusoidal involution of the cliff would never allow a smooth transition from the upper to the lower state; no smooth gradient of descent, in real terms. So there had to be discontinuity between the top and the bottom lines of the S he'd drawn—an abrupt flip from State A to State B; and that was, mathematically speaking, a "catastrophe."

(There is no gravity in dreams . . .)

He waved a cigarette at his colleagues: Mary Strope, looking bewildered but defiant; Oliver Hart, wearing a supercilious expression; senior consultant Daniel Geraghty, looking frankly outraged.

"Taking the problem in its simplest mathematical form, is this a fair representation of the onset of cancer?" Adrian demanded. "This abrupt discontinuity, here? Where we fall off the cliff—"

Rapping the blackboard, he tumbled Gitane ash and chalk dust down the cliff. The obsession with this particular brand had taken hold of him even before his trip to France, and he'd borrowed so many packs from the smoking room downstairs (where a machine was busily puffing the fumes from a whole range of cigarettes into rats' lungs) that Dr. Geraghty complained he was sabotaging the tests and Oliver Hart suggested flippantly that Adrian should be sent to France *tout de suite,* Thibaud-wards, if only to satisfy his new craving . . .

"I suggest that, instead of a progressive gradient of insult to our metabolism, we abruptly flip from one mode to the other: from normal to malignant. Which is perfectly explicable, and predictable, using catastrophe theory. Now the immune system shares one major formal similarity with the nervous system. It too observes and memorizes events. So if we view the mind—the superior system—as a mathematical network, could it predict the onset of cancer mathematically, *before* we reach the stage of an actual cellular event, from this catastrophe curve? I believe so."

He swiveled his fist abruptly so that the stick of chalk touched the blackboard rather than the cigarette. Yet it still looked like the same white tube. Then he brought the chalk tip screeching from the cliff edge down to the valley floor.

Their eyes saw the soft cigarette make that squeal—a scream of softness. Adrian smiled, as his audience winced in surprise.

"But how can the mind voice its suspicions? I suggest in dreams. What are dreams for, after all?"

"Data processing," replied Oliver Hart impatiently. "Sorting information from the day's events. Seeing if the basic programs need modifying. That's generally agreed—"

"Ah, but Thibaud believes they are more."

Oliver Hart was dressed in a brash green suit; to Adrian he appeared not verdant and healthy, but coated in pond slime.

"For example, to quote my own case, I am approaching a cancer—"

Deftly, with sleight of hand, Adrian slid the cigarette off the cliff edge this time, amused to see how his three listeners braced themselves for a repeat squeal, and shuddered when it didn't come.

"I shall have the posterior pons brain area removed in an operation. Then I can act out my dreams as the slope steepens towards catastrophe—"

Mary Strope caught her breath. She stared, horrified.

"Enough of this rubbish, man!" barked Geraghty. "If this is the effect Thibaud's notions have on you, I can only say your visit there was a disaster for the unit. Would you kindly explain what twisted logic leads you to want part of your brain cut out like one of his damn cats? If you can!"

"If I can . . . No, I couldn't have it done in France itself," reflected Adrian obliquely. "Probably it'll have to be in Tangier. The laws are slacker there. Thibaud will see to the arrangements . . ."

Mary half rose, as though to beat sense into Adrian, then sank back helplessly and began crying, as Geraghty bellowed:

"This is a disgrace! Don't you understand what you're talking about any more, man? With that part of the brain destroyed there'd be no cutoff in signals to the muscles during your dreams. You'd be the zombie of them! Sleepwalking may be some temporary malfunction of the pons—well, sleepwalking would be nothing to the aftermath of such an operation! Frankly, I don't for one instant believe Thibaud would dare carry it out on a human being. That you even imagine he would is a sorry reflection on your state of mind! Stop sniveling, Mary!"

"Adrian's been overworking," whimpered Mary apologetically, as though she was to blame for his breakdown, whereas she had only been offering love, sympathy, comradeship.

"Then he shall be suspended, *pro tem*. D'you hear that, Rosen? No more waltzing off to France, making fools of us."

"But I shan't be living long," Adrian said simply. "You forget the cancer—"

"So there, we have located it," Jean-Luc Thibaud had declared proudly, "the mechanism that stops nerve signals from the dream state being passed on as commands to the body. Essentially the pons is a binary switching device. The anterior part signals that dreams may now take place, while the posterior part blocks off dream signals to the muscles . . ."

Thibaud seemed a merry, pleasant enough fellow, with a twinkle in his eye and the habit of raising his index finger to rub the side of his nose, as though bidding for cattle at some country auction. His father was a farmer, Adrian remembered him saying. And now his son farmed cats, not cattle.

"Thus we can remain relatively limp during our nightly dance with the instinctual genotype which psychologists so maladeptly label the unconscious mind . . ."

A hall of cats.

Each cat was confined in its own spacious pen, the floor marked off by a bold grid of black lines like graph paper. Lenses peered down, recording every movement the animals made on video tape.

Most cats were asleep, their eyes closed.

Most cats were also on the move. Scratching. Spitting. Arching their backs. Lapping the floor. Fleeing. Acting out their dreams in blind mute ritual dances of flight, rage, hunger, sexuality . . .

And a few, a very few, were only dozing, not dreaming. These didn't move. They hadn't drifted far enough down the sleep gradient yet. Soon they too would rise, and pace, and fight. Soon they too would lap the floor and flee. Till they dreamed themselves to death, from sheer exhaustion. It was tiring work, dreaming, down on Thibaud's cat farm.

From each cat's shaven skull a sheaf of wires extended to a hypermobile arm, lightly balanced as any stereo pickup, relaying the electrical rhythms of the brain to be matched against this dream ballet taped by the video machines.

"And still I am dissatisfied, M'sieur Rosen! Still, we see only the genetic messages for the most basic activities being reinforced. That's what this is, you realize? A genetic reinforcement. Errors creep in from one cell generation to the next. Too many errors, and—pouf! An error catastrophe. Death. So dreams strive to reinforce the purity of the genotype—like the athlete trying to keep himself fit by exercises. Dreams are error correction tapes manufactured out of each day's new experiences. But gradually we begin to dream out of the past, as the years go by. Increasingly we scavenge yesteryear. Soon, we are scavenging yesteryear's dreams themselves—using bygone, frayed correction tapes. We lose the capacity to make new ones. We dream vividly of childhood and it seems we are re-entering paradise as we sleep. Alas, that's all too true. We're about to leave the world, literally—for the cold clay of the cemetery."

"Yet I wonder, Dr. Thibaud, what if error is an essential part of our life process? What if, in order to be able to grow, we must also be able to die?"

"Yes, indeed—the cruel dialectic of nature!"

"Well then, what part has the cancer cell to play? It's the only truly immortal cell. It alone copies itself perfectly, without any error. And it kills us by doing so."

"The difference between cell replication and cell differentiation is a knife edge we must all balance on, M'sieur Rosen."

"Yet we all have cancer, potentially. Viral cancer lurks in everyone's cells in a latent form, did you know? I want to know why. Doctors perpetually set themselves up to cure cancer—to cure polio, to cure everything else they label as disease. And that's supposed to be the whole work of medicine. But how many doctors ever trouble to glance at the whole system of life and evolution that a 'disease' functions in? None whatever!"

A cat—a mangy, skinny alley tabby—pounced on the invisible prey that it had been wriggling its way towards all the time they talked. But almost at once it leaped away again. Its fur stood on end, its tail bushed out, as it backed cowering into the corner of its pen.

"Did you see that, M'sieur Rosen? You could call it a catastrophe, in your terms—that sudden switch from fight to flight. The mouse becoming a monster in its mind. Yet how much do we really see? It's as you say about medicine—scratching the surface. Examining the arc of the circle and thinking that's all there is to the figure! But the inner landscape of the dream must be just as important as the actions. If not more so! In fact, I'm inclined to think the full subtleties of the genotype can only be coded into the dream as environment. Yet how to show that? Still, we're only starting on our journey inwards. Come, see the darkroom. We raise some other cats in black light and isolation from birth, so that they display the perfect archetype of a dream . . ."

The alley tabby awoke, as they retraced their steps, and whined from the fretful exhaustion of having slept.

"Presumably, in an archetypal setting . . ."

* * *

Their glider bounded over the turf as the winch driver heeded the blinking of the aldis lamp from the control caravan, then slid smoothly into the air, climbing gently towards and upward of the winch gear. Mary pulled back softly on the stick, increasing the angle of climb to balance the downpull of the cable, till at eighty degrees to the winch and an altitude of a thousand feet she dipped the nose briefly, pushed the cable release knob, then climbed away.

"What if it doesn't let go?"

"It disconnects automatically, if you're at a right angle to the winch—which you shouldn't let happen."

"It could jam."

"There's a weak link in the chain, Adrian. By design. It's fail-safe." But she sounded exasperated.

The hill up-current sent the glider climbing towards scattered woolly cumulus in a sky which was the blue of a pack of French cigarettes, as Mary manipulated the controls efficiently, banking, centralizing, and taking off rudder, then repeating the same turning maneuver with a minimum of slip and skid. And so they spiraled aloft.

Her hair blazed back in the wind when the glider did slip to the right briefly on one turn, uncovering the firm rhombus of her cheekbones, and a number of small brown moles just in front of the hairline. For a redhead her skin was only lightly freckled. It resembled the grain of an old photograph more than distinct freckles. Adrian loved touching and stroking those few hidden blemishes when they were in bed together, but it generally took a strong wind to whip the bonfire of hair back from them.

"So you're set on going to France?" she said at last. "I think Geraghty would rather Oliver went."

"Oliver doesn't have my special interest."

"What interest? It's nonsense!"

"You know very well."

"I know nothing of the sort! You're perfectly healthy. Why else do you refuse to take a medical? It would show how wrong you are."

"I can . . . examine myself. The dreams, you see. It would spoil everything to have some silly checkup. It would ruin the experience. I must keep perfectly clear and neutral."

The glider skidded badly then, as Mary angrily used a bootful of rudder, and the nose began hunting, pitching to and fro.

"You realize you're wrecking our relationship? Your scientific credibility too! If that matters to you!"

"My dreams have a shape to them. I have to . . . live them out."

Correcting the trim of the machine, Mary spiraled the glider through the woolpack, avoiding entering cloud. They soared above the snow cocoons into open sky; the clouds swept by below them now like detergent froth on rivers of the air—the vale and downs being the soft clefted base of this surge of translucent streams. They continued a stable upward helix for another few hundred feet till uplift weakened and Mary swung the machine away towards a thermal bubble on which another pilot was rising a mile away, in company with dark specks of swifts and swallows catching the insects borne up along with the air.

But if they'd entered cloud, reflected Adrian, and if another pilot had also done so, and the curves of the two gliders intersected in the woolly fog, then there'd have been . . . discontinuity: a catastrophe curve.

Marguerite Ponty accepted the infrared goggles back from Thibaud and Rosen to hang on the hook outside the second of two doors labeled Défense d'éclairage!

The slim woman's dark glossy eyes were heavily accented by violet eye shadow which made huge pools of them; as though, having spent too many hours in null-light conditions tending to the darkroom cats, her senses were starting to adapt.

Her hair was short and spiky, gamine style. She wore dirty plimsolls, blue jeans, and a raggy sweater under her white labcoat, the loops of the knitting pulled and unraveled by cats' claws. From her ears hung magnificent golden Aztec pyramids of earrings. Her scent was a strange mix of patchouli and cat urine: clotted sweetness and gruelly tartness grating piquantly together.

"The pons area is lesioned at one year old," Thibaud commented. "They've never seen anything. Never met any other cat but their mother. Yet in their dreams they prowl the same basic genetic landscape. The computer tells us how they show the same choreography—only purified, abstracted. What is it, I wonder? A Paul Klee universe? A Kandinsky cosmos? Has anyone unwittingly painted the genetic ikons?"

"Let's hope not Mondrian," laughed Marguerite. "What a bore!"

"Blind people dream," Rosen reminded him. "Surely they don't visualize. They smell, they hear, they touch."

"And out of this construct their landscape, yes. Same thing. It's the putting together that matters. The shaping."

"Topology."

"Exactly. I was only using a metaphor. Let me use another: our blacklight cats are dancing to the same tune as our sighted tribe. Yet they experience next to nothing in their lives."

Rosen couldn't help glancing pointedly at Marguerite Ponty's looped and ragged sweater. They experienced her.

"Which proves that dreams are control tapes for the genes, not ways of processing our daily lives. But come. It is time to show you our cancer ward. We use nitrosoethylurea to induce tumors of the nervous system—thus the immune battle is fought out within the memory network itself! The basic instinctive drives yield right of way to a more urgent metabolic problem. You'll see the shape of catastrophe danced. That's what you came for."

Rosen grinned.

"Immune dreams, yes. But what landscape do they dance them in?"

"Ah, there you ask the vital question."

Another day. Another flight. Another landing. And Rosen had been to France, by now.

Mary pointed the glider down steeply towards the two giant chalk arrows cut in the field.

It struck him that she was diving too steeply; but not so, apparently, for she raised the nose smartly to bring them out of the dive flying level a few feet above the ground, the first arrow passing underneath them, then the second. They slowed as she closed the air brakes, pulling the stick right back to keep the nose level, till they practically hovered to a touchdown so perfect that there was no perceptible transition between sky and ground. She threw the air brakes fully open, and they were simply stationary.

Cursorily she rearranged her hair.

"Nature's so bloody conservative," Adrian persisted. "It has to be, damn it, or there wouldn't be any nature! You can't have constant random mutations of the genotype. Or you'd always be losing on the swings what you won on the roundabouts. So once a particular coding gets fixed, it's locked rigidly in place. All the code shifts that have led from the first cells through to cabbages and kings, have operated upon redundant DNA, not the main genome. Look around, Mary. How diverse it all seems! Sheep. Grass. Birds, insects, ourselves. So much variety. Yet genetically speaking it's almost an illusion. Quality control is too strict for it to be any other way. Just think of the Histone IV gene for DNA protein binding. That's undergone hardly any change since people and vegetables had a common ancestor a billion and a half years ago. Biological conservatism, that's the trick! But what's the most conservative cell we know?"

"Cancer, I suppose," admitted Mary. "What are you driving at now?"

"Quality control to the nth degree!" he rhapsodized. "That's cancer. And now we know there's viral cancer lying latent in everyone's cells. It's part of our genetic inheritance. Why, I ask you?"

"To warn the immune system," Mary replied brightly. "When a cell goes cancerous, the virus has a chance to show its true colors as an alien. Our immune system couldn't possibly recognize cancer as hostile tissue otherwise."

"Very plausible! Then why's the system so damned inefficient, if we've got these built-in alarms? Why do so many people still die of cancer? Have medical researchers ever asked that, eh? Of course not! They never think about the whole system of life, only about correcting its supposed flaws."

"Maybe more cancers get stopped early on than we realize?"

Adrian laughed.

"So you think we may be having low-level cancer attacks all the time—as often as we catch a cold? There's an idea! But I fancy that viral cancer's not locked up in our cells to warn the immune system at all. The reason's quite different. And it's so obvious I'm surprised no one's thought of it. Cancer's there to control the quality of replication of the genotype—because cancer's the perfect replicator."

"That's preposterous!"

"Cancer isn't the alien enemy we think. It's an old, old friend. Part of the Grand Conservative Administration presiding over our whole genetic inheritance, keeping it intact! It's a bloody-minded administration, I'll grant you that. It has to be, to keep in power for a billion years and more. Thibaud was fascinated when I outlined my theory. It casts a whole new light on his genetic dream idea—particularly on the class we're calling 'immune dreams.' Cancer's catastrophe for the individual, right enough. But for the species it's the staff of life."

"Your health," Thibaud grinned broadly: a farmer clinching a cattle deal. Marguerite Ponty smiled more dryly as she raised her glass, clicking her fingernails against it in lieu of touching glasses. Her earpieces shone in the neon light, priestesslike. Were they genuine gold? Probably. Her joke about Mondrian referred to her father's private collection, it transpired. Rich bourgeois gamine that she was, she'd chosen the role of a latter-day Madame Curie of the dream lab—as someone else might have become a Party member, rather than a partygoer. There was something cruelly self-centered in the way she regarded Rosen now. Of the two, Thibaud was much more vulgarly persuasive . . .

Thibaud also looked genuinely embarrassed about the wording of his toast when the words caught up with him.

"A figure of speech," he mumbled. "Sorry."

"It doesn't matter," said Rosen. "It's the logic of life—the cruel dialectic, as you say. Thesis: gene fixation. Antithesis: gene diversification. Synthesis: *ma santé*—the sanity of my body, my cancer."

"Yours is such a remarkable offer," Thibaud blustered, beaming absurd, anxious goodwill. "You say that an English specialist has already confirmed your condition?"

"Of course," Rosen produced the case notes and passed them over. He'd experienced no difficulty forging them. It was his field, after all. And if Thibaud suspects anything odd, thought Adrian, odds are he's only too willing to be fooled . . .

Still, Thibaud spent an unconscionably long time studying the file; till Marguerite Ponty began flicking her gold earpieces impatiently, and tapping her foot. Then Rosen understood who had paid for much of Thibaud's video-tape equipment and computer time. He and the woman regarded one another briefly, eye to eye, knowingly and ruthlessly. Finally, hesitantly, Thibaud raised the subject of the clinic in Morocco.

"It will take a little time to arrange. Are you sure you have time—to revisit England before you come back here?"

"Certainly," nodded Rosen. "I need to explain some more details of the theory to my colleagues. The cancer isn't terminal yet. I have at least two months . . ."

They sat in Mary's convertible, watching other gliders being winched into the air: close enough to receive a friendly wave from one of the pilots, with whom Mary had been out to dinner lately. A surveyor or estate agent or some such. Adrian hadn't paid much attention when she told him.

Or was he a chartered accountant?

The winch hummed like a swarm of bees, tugging the man up and launching him over the vale. Geologically speaking you'd classify it as a "mature" valley. In a few more tens of thousands of years, weather action would have mellowed it beyond the point where gliders could usefully take advantage of its contours. But at this point in time there was still a well-defined edge: enough to cut the vale off from the hill, discontinuing, then resuming as the landscape below.

Mary lounged in the passenger seat. She was letting Adrian drive the car today. It was the least she could do, to show some residual confidence—since Geraghty's suspension of him; though it was some while since they'd actually been up in a glider together.

Softly, without her noticing, he reached down and released the handbrake.

Once she realized the car was moving of its own volition towards the edge, he trapped her hands and held them.

"Look," he whispered urgently, "the genetic landscape."

"Adrian! This isn't a dream, you fool. You aren't asleep!"

"That's what they always say in dreams, Mary."

He pinned her back in her seat quite easily with dreamlike elastic strength while she cursed and fought him—plainly a dream creature.

Soon the ground leapt away from the car's tires; and he could twist round to stare back at the face of the hill.

As he'd suspected, it betrayed the infolded overhang of catastrophe. The shape of a letter S. Naturally no one could freewheel down such a hill. . . .

Later, he woke briefly in hospital, his head turbaned in bandages, as seemed only reasonable after an operation to excise the posterior pons area of the brain. He found himself hooked up to rather more equipment than he'd bargained for: catheters, intravenous tubing, wires and gauges proliferating wildly round him.

He stared at all this surgical paraphernalia, curiously paralyzed. Funny that he couldn't seem to move any part of him.

The nurse sitting by his bedside had jet-black hair, brown skin, dark eyes. He couldn't see her nose and mouth properly—a yashmaklike mask hid the lower part of her face. She was obviously an Arab girl. What else?

He shut his eyes again, and found himself dreaming: of scrambling up a cliff face only to slide down again from the overhang. Scrambling and sliding. A spider in a brandy glass.

KATHRYN CRAMER

Forbidden Knowledge

Down in the grope room of the Science with Gargoyles bookstore, Phelony Verfall found the old Ambassador. While meditating upon the vanishing of Ext and its multileveled meanings, she breathed slowly and deliberately, striving for a state of utter deadly calm—imagining herself as the eye of a particularly violent hurricane—walking into the store and past the glass cases at the front counter, creeping down two narrow flights of stairs from which someone had taken up the stair carpet and left the heads of nails sticking out three centimeters, going into the room where, according to Reformist informers, the Ambassador now spent most of his waking hours. After resigning from his lifetime appointment, Ambassador Winston Otto Notsniw had come here, bringing with him an antique pilot's seat salvaged from a Boeing 797 jet and a set of morally instructive photographs. Phelony was not at all eager to visit him, but the Reformist Executive Committee had decreed that because of her superior IQ, 167 plus or minus three points for experimental error, and because of her good looks—although her hair was mousy brown and cut too short, she had a lovely, rounded bottom—the job was hers.

The Ambassador sat all day long every day, withered and puppetlike, in the grope room, which he had made his home. His only visitor until now had been the venal store owner and the store's customers who came to the room, overcome by intellectual experience, seeking release and fulfillment. The room had been built to look like a grotto. The walls were fiberglass, airbrushed to look like stone. With the Ambassador's arrival it became a sort of grotesque throne room. He had paid the owner to install a picture window so he could sit all day and look out over Evanthia Bay. He had also required that his collection of photographs be hung on the walls. For half a moment Phelony felt as though she'd entered the hall of the mountain king.

Phelony was here despite frustrating weeks explaining repeatedly to the Reformist Executive Committee—of which her current lover, Henley Hornbrook, was a member—why she shouldn't be required to do this. The Ambassador had no idea who she was, and was therefore unlikely to listen to her opinions on the legalities of Ext. Phelony was shy and had a tendency to lecture to people she didn't know when in situations that required her to talk to them. (That had been the committee's reason for firing her from her paid position as Volunteer Coordinator; why didn't they care about that now?) If the Ambassador had resigned and now spent all his time in an obscure bookstore, what was the likelihood that he would come out of his hermitage to lobby on the Reformists' behalf, especially since he had been the primary advocate for imprisonment and execution of their electoral candidates in the Mood Vision scandal five years before? But when she realized that the committee wasn't listening to her, that it must have all been Henley's idea, she gave in.

Phelony and Henley had had a nearly terminal fight the night before the committee voted to assign her the job. He had been lecturing to her for weeks about crystal lattice techniques for modeling oppressive systems when she went to the library and read up on the subject so she would be able to hold up her end of the conversation. She brought up the subject on that particular evening and as soon as she began to argue with him he refused to discuss crystal lattice techniques or anything else and turned on the holovision.

When Phelony first moved in with him, they spent all their time arguing and having loud sex (so loud that their neighbors frequently banged on the wall in protest), but as Henley rose within the Reformist heirarchy and as their relationship matured, increasingly they spent their time together watching holovision reruns of popular executions. On that particular night, angry and frustrated with trying to please him, Phelony stood in the middle of the hologram, fists clenched, screaming at the top of her lungs that she wouldn't move from in front of the electric chair until he listened to what she had to say about the oppressive aspects of their relationship. Henley had always been the kind to take input and feedback over suggestions and wise advice. Since nothing she ever had to say came through the proper channels, he had learned to ignore her until she got it right. This much he had explained to her on one of those rare occasions when she forced him to talk about his feelings. That particular night he sat quietly until she was done screaming and then said, "Trust is good; control is better," and went off to bed. The Executive Committee meeting the next morning was held in closed session, so she wasn't allowed in. But she was sure Henley was behind all this.

And now Phelony stood before the Ambassador, armed only with a deeply held ideological and personal commitment to guilt, a profound belief in her own worthlessness, and a rap—a canned speech—which she had carefully memorized before coming to argue with the Ambassador about Ext. For her, Ext had a personal and intimate meaning that she had never dared discuss with anyone. Ext was a form of fear. It was fear within a relationship that, if the relationship is to survive, one must cause to vanish. The Executive Committee had assigned her to persuade Ambassador Notsniw, by whatever means necessary, to denounce publicly the Notsniw Law, the Ambassador's brainchild, what holovision editorialists called his legislative masterpiece. Editorialists said it was the law that would protect the public from forbidden knowledge from the Worm Planet, protect the public from perverse concepts that creep into the brain, which turn ordinarily social human beings into something else entirely. The law made publication or possession of materials pertaining to Ext a federal offense, carrying penalties of fifty thousand Erdmarks or ten years in prison without hope of parole. The mathematical techniques for gaining access to the powers of Ext had been learned by xenobiologists while studying the Group Mind on the Worm Planet and were now in use by three-quarters of the human population in both inhabited solar systems as a method of eliminating emotional problems by translating them into the external world. The law was therefore in effect in both inhabited solar systems. The pop-psychology tape which popularized Ext was based upon the case of a little boy named Winston Notsniw, son of the cook on the xenobiological expedition, who was miraculously cured of his autism when the Group Mind revealed to him the geometricity of coiled gastropod shells. Ext immediately became a cultural obsession and the little boy grew up to become Ambassador to the aliens. The Notsniw Law was a masterpiece because it legally defined Ext, which until that time had remained in the same linguistic limbo as obscenity.

After Phelony had stood staring at him for an uncomfortably long moment, wishing she could melt into one of the fiberglass walls, the Ambassador waved her to be seated with his deformed right hand which, curiously, lacked an index finger all the way down to the third knuckle. She sat down on a block of smelly, discolored foam rubber which had been cut to look like an outcrop of rock. He poured her a steaming snifter of the Black Milk of Morning, which she accepted, sipping cautiously. Because she had not expected to be received hospitably, she was taken by surprise. Grinning widely, his first expression since she entered the room, he pushed a platter of jumbo shrimp, perched atop thirty hardcover copies of *Equations Out of the Underbrain,* five and one-half centimeters in her direction.

"I dine on rats," he said; his facial expression vanished quite slowly, beginning with the deep creases in his warty forehead and ending with the grin which remained for some time after all other expression had gone.

She smiled back blandly, unsure how to react. He didn't seem to be joking, nor did he seem sincere. His flat affect unnerved her so much that for one tetrahedral moment she wanted to flee; three steps to the door, up two flights of rickety stairs, down the dimly lit main aisle of the bookstore, and she would be away. But instead she turned her eyes toward the picture window and looked out upon Evanthia Bay. The picture window seemed incongruous in a grope room, but she supposed that was the real reason why he'd had it installed. And outside was the external world with all its unnatural geometricity. She heard the voices of birds and the constant roar of waves and the whirr of machines as dust-speck men peeled the metallic skin off what appeared to be a giant golden onion out on the tide flats. The waves were composed of gray-green wavelets and the coastline between Phelony and their cramped studio apartment where Henley waited for her (ah, Henley of the long green braids!) on the other side was infinitely long because it detoured into every inlet and around every rock and grain of sand, so she knew she could never go home.

"Hi! My name is . . ." she began. But then she stopped. He didn't seem to care who she was, didn't care that she was twenty-five years old, that she had had five wisdom teeth instead of the ordinary four, didn't care that she had made a well-reasoned political decision never to shave her legs, didn't care that when her parents had first brought her baby brother home from the hospital she put her favorite book in his crib and said "baby read book," didn't care that she had pulled the legs off spiders when she was five, didn't care that although she loved Henley she didn't think she could stand him much longer. Indeed she had an intense feeling that this man didn't care at all, as though he had developed not caring to a high art. All he had to do was sit there in his chair, and she could tell that he didn't care about her or anything she might have to say. She might as well go home because he wasn't going to listen. But she felt obligated to the committee at least to give it a try. So she tried.

"Causality as we understand it," she said. "is derived from the assignment of objects into categories. Categorization, in turn, is derived from the psychology of the individual. Thus causality is derived from psychology. This is the mechanism which gives Ext its power. But if Ext is defined by law, and the law imposes its metaphor upon our collective psyche—the collective psyche being the glue which holds human civilization together—then causality itself becomes static and civilization collapses. Therefore the Notsniw Law must be rescinded before it brings about the collapse of civilization as we know it."

As she paused, waiting to see if her opinions would elicit some sort of response, she felt sorry for him, perhaps because his olivine shirt had frayed cuffs and was shedding sugary green crystals on the rough wooden floor. He looked terribly small and pathetic in person. In his lap was a green hoop about the size of a dinner plate which was, she assumed, a symmetry multiplier, a souvenir from his career among the aliens; stretched across the hoop and attached with odd-shaped copper staples was a milky and translucent membraneous diaphragm. She had expected the Ambassador to be much more imposing and demonic, but instead he had the air of a pet monkey abandoned by its owner. The alien artifact, however, was unsettling.

His only response to her speech was to drop a shrimp onto the diaphragm, upon which images of the shrimp appeared in a four-leaf-clover pattern, sending a fine powder of what used to be shrimp breading sifting down into his lap. The symmetry multiplier came from the Worm Planet, an odd name for an odd place, called the Worm Planet both because the larval stage of its Group Mind resembled giant opalescent worms and because it was the planet where the first Worm Hole was discovered. The Group Mind used the symmetry multiplier to teach its larva about geometry. This particular hoop demonstrated one of the plane's seventeen symmetry groups, the one Phelony knew as the p4m group. The hoop peeled off a thin layer of matter and portioned it into images.

Tossing the symmetry multiplier, which she caught deftly by the rim with her left hand, he said, "No devil sees doom, nor star era radar a ten megaton." His ferociously green eyes seemed focused on the tip of her nose. She nodded gravely in agreement, trying to look as though she understood. Perhaps, she thought, he is making some broad observation about humanity's lack of foresight which I can connect up with attempts to legally define Ext.

She took his comments to be a conversational opening and continued with her rehearsed speech: "What you must understand, Ambassador, is that because public commentary on Ext has been left entirely to elderly court judges and publicity-seeking politicians—the sort of person you yourself used to be—who are concerned with legality rather than reality, and to up-tight private citizens who hate Ext enough to spend their evenings, weekends, and holidays campaigning to have Ext banned, those inconsiderates who intend that Ext be restricted by force of law—thereby diminishing the necessary flexibility of the nature of reality for the vast majority of the human population—present the public, considered both as a collection of individuals and as one gigantic whole, with no reasonable alternative method of exorcising the chaotic emotions. Because, lacking diversity of perception, individuals will act entirely in their own best interests, thereby causing all of humanity to descend into barbarism, diversity of perception is an ingredient essential to the very existence of democratic society, and therefore because a legally established definition of Ext will, in the long run, subvert the very structure of civilization, the motivation behind the Reformist outreach to you, the motivation behind this visit, is to get you to denounce all further attempts to legally define Ext and to advocate the revocation of your own part in this disaster—the Notsniw Law. If civilization is to have any chance of survival, any chance at all, all attempts to define Ext must cease and the Notsniw Law must be revoked."

She looked through the curl of steam rising in the room's chilly air, to the glass cradled in her right hand, waiting for his response. The moment seemed too precise. Not larger than life, but sharper. Looking at the steam, she dimly perceived its shape: how each eddy was composed of smaller eddies, which in turn were composed of smaller eddies, and so on until there were only H_2O molecules wriggling randomly in the air before her. The Black Milk of Morning had left her mind almost painfully sharp, although her teeth were probably stained midnight blue by now just like the Ambassador's. There was a ring of bubbles where the Black Milk met the glass, each bubble squeezed between smaller bubbles. Where the glass was chipped and cracked, the spider's webbing of cracks was stained blue-brown. She looked closely at the jagged place on the lip and squinted.

"Reviled yam, flow!" said the Ambassador, pouring himself another snifter of Black Milk. For a moment what he said made no sense to her, but she assumed, although she didn't know for sure, that the Black Milk must be made of fermented yams.

The sharp edge of her glass became mountains, which in turn had foothills. "Ah," she said. "Self-similarity in solid, liquid, and vapor. When we try we can see Ext vanishing everywhere, Ambassador. We need only try. In return we get release and fulfillment far beyond any experienced by a human being before Ext. Would you really let this be taken away? Or will you denounce the Notsniw Law?" Phelony had always found it miraculous how her very deepest secrets, the ones she hid even from herself, were translated by Ext into the structure of the external world. As she smiled at him, trying to re-establish their rapport, she thought of all those hours spent tabling by the ticket desk in the space port, spent chairing meetings, spent programming volunteers, spent networking with sympathetic organizations, spent leafleting the vitreous streets of Zwerin, and she thought that if he would denounce the Notsniw Law, then he would make it all worthwhile.

Although she gobbled down five jumbo shrimp, after setting her glass down between her thighs, to show him that she accepted his hospitality, her efforts had no noticeable effect upon him. A few minutes later, when the Ambassador leaned forward unexpectedly, Phelony drew back.

"Emit noon time!" he said in a commanding tone. Phelony looked at her watch, which read 5:47 P.M. Noon, she wondered. What could he mean? He waved his finger stump at her and said, "Wolf may deliver not a gem." She tried to make sense of what he said. Was he referring to himself as the wolf? And what did this have to do with noon? If he was the wolf, what was it that he was or was not going to deliver?

At this moment, for the first time, she realized she was frightened of him. Inside her heavy shoes she curled and uncurled her toes. She was alone in the room with this famous man—the first and only Ambassador to the Group Mind, for whom no replacement had yet been found—and he was crazy. This contradicted what Henley had explained to her in such great detail: that after two decades of public service during which the Ambassador, a sincere and altruistic man, worked to restrict public access to information about Ext despite the fact that he himself had been the first human to be helped, even cured, by it, and thus depriving literally billions of people of the benefits he himself had enjoyed, the Ambassador had finally been forced by guilt to resign. But no. The Ambassador was crazy. That was all. Sometimes the simplest explanation is the best.

Phelony was tensing up, preparing to rise from the foam block and run from the room, when the door burst open. In slunk a tall young man with curly red hair starting wildly from his head. Phelony assumed he must be a mathematician because he had one hand shoved down his pants and in the other he held the archaic text, the historical and neo-religious basis for the name of Ext, Irving Kaplansky's *Fields and Rings,* second edition. He hid himself in a corner of the room, concealed by a nook of fake rock.

"Are rats radar Ron?" the Ambassador asked him. But a groan was the young man's only response. The Ambassador grinned and nodded, saying, "Mood sees."

Phelony had no more idea how to reply than if he had been speaking a foreign language. She munched on jumbo shrimp and contemplated her escape—wondering where she could go that the Reformists would never find her or alternatively what she could tell them that would explain her departure with the job still undone—and she stared at a photograph on the south wall. When the photograph caught her eye, she thought it was a well-composed portrait of a sleeping woman wearing a jumper, the kind of jumper Phelony might have worn before she became a Reformist and realized that it was wrong, the kind with buttons down the front. Her next thought was, this can't be real. Those weren't the shoulder straps of a jumper. Nothing was what it seemed. Rather the flesh had been peeled back from the woman's shoulders in neat little strips. The shoulder incisions allowed her breasts to be pushed back off the fat on her rib cage, forming humps that looked like withered apples. Those weren't buttons. The graininess in the center of each was bone marrow. That was the cross-section of her rib cage.

Hair arranged just so, eyes closed as though she were asleep, the lumpy texture of exposed fat contrasting with textures of hair, bone, and smooth skin, the handwritten card beneath the picture read, "Ext death, age 26." Phelony felt as though she too were going insane. Nothing seemed real. It was a photograph of a young woman just after autopsy, a young woman who looked familiar, perhaps because Phelony had seen her on holovision, but more likely because she had been a Reformist.

"Lived on star," said the Ambassador. As the sun set outside the window, Phelony was on the verge of screaming. But she either had to get away or go on with her speech. The man with the book had reminded her of the really crucial part.

"We shall have no need to assign meaning to Ext itself," she said. "We shall speak only of its vanishing." The Ambassador's face became molten, a boiling sea of anger.

"No Enid, I . . ." he said. As his voice trailed off, his expression cooled, crystalizing into his usual, placid nonexpression. The man was an existential Cheshire cat; he always vanished back into the gloom of nonbeing after every display of emotion. For a moment she thought he had mistaken her for someone else, someone named Enid.

Sharp as shattered glass slicing skin of the hand that holds it, her understanding cut quickly to the problem's core and drew blood: "no Enid I" backwards is "I dine on." The cycle was complete and something dreadful was about to happen. In the air around her was the fear, but as the moment lasted she felt less and less. She was using the power of Ext to do battle with the fear, using Ext to hide the fear.

She threw it at him, the thing in her hand. Suppressed violence always escapes its dungeon. She heard him scream. Because the screaming scared her, she leaped up from her chair, knocking over the jumbo shrimp and the stack of math books and shattering her glass on the floor. It began as a man's scream and rose in pitch by abrupt jumps. Phelony had never in her life heard anyone scream quite that way. The scream went silent when the Ambassador fell out of the pilot's seat onto the unfinished wooden floor. She tasted blood. She had bitten a hole in her tongue. She could hear the man in the corner moaning with pleasure, and the voices of birds and the constant roar of the waves and the clicking and whirring of machines, oh yes, the clicking and whirring of machines! In the pink glow of twilight, filtering in through the window, she saw eight heads with faces all the same. She counted them: two-four-six-eight. Four pairs like the leaves of a four-leaf clover. It had eight long thin heads. The face from holovision documentaries, from political conventions where she had been sent to spy, from advertising billboards, was now surrealistically and symmetrically superimposed upon each of them; every head had former Ambassador Winston Otto Notsniw's face.

Phelony screamed as the heads began to droop. She screamed as the heads became like axle grease or grape jelly. And when the head jelly became a pool of dark liquid, deprived of all its long-chain molecules, and flowed into Phelony's spilled Black Milk, she was still screaming. Around the Ambassador's neck was the hoop of the Worm-thing, the symmetry multiplier, with shreds of safety membrane still attached. Phelony Verfall had killed him.

She felt a warm and comforting hand upon her shoulder and for a moment the very shape of smoke was revealed to her. Indeed she felt a piece of time missing from the moment. She felt calm, reassured, almost confident, as though the events of the past hour were only a prelude to something wonderful. The man with the red hair stood beside her, naked and glowing. From the neck down, all of his bodily hair was pale green, just like Henley's. Suspicious, she pulled on a lock of wild red hair. A red curly wig came away in her hand, exposing long green braids, looped around his head and held in place by bobby pins. It was Henley. He had come to watch. In an effort to console her, he held out his copy of the archaic text, opened to page 187 where it was written: "*13. The Vanishing of Ext.* We shall have no need to assign meaning to Ext itself; we shall speak only of its vanishing."

Henley, Henley. Henley had been in the room during the crucial time, and he had tried neither to help her nor stop her. And now he was smiling. She would hurt him.

The text was precise and accurate, but not in any sense Henley would understand. But now Phelony understood. She understood the quotation's shape, texture, found meaning preserved in it for her like a woolly mammoth encased in glacial ice. He caressed her cheek smiling luminously, condescendingly. As he took the book from her hands, she knew the future would be very different. With only momentary effort, applied through Ext, she made him vanish.

It had all been intended, not by the Reformists, but by the aliens. She must go to the Worm Planet immediately: Phelony Verfall, the new Ambassador. She must hide the forbidden knowledge of the Worm Planet.

GEORGE ZEBROWSKI____.

Gödel's Doom

"So what are you going to beg time for now?" I asked as Witter slid in across from me in the cafeteria booth. A thin, hyper type, he folded his hands in front of my coffee and said, "It's an experiment I want to run on the new AI-5." He spoke very precisely, very insistently, as usual. "I've been haunted by it all my life, but now it can actually be done."

"What do you mean?" I asked, picking up my coffee, afraid that he would knock it over.

"Well, previous Artificial Intelligences were too slow and not capable of complex inference. The question is how much time can you give me?" He brushed back his messy brown hair.

"How much do you need?" I sipped the coffee, sensing his restrained excitement. Witter had always been a valuable worker, so I had to listen and try to keep him happy, within reason, despite his nervous enthusiasms. But he was never satisfied with merely testing equipment and programs for industrial applications.

"I don't know," he said cautiously. "A lot maybe. More than a couple of days."

I put down my coffee, irritated. "You don't know? Can't you estimate?"

"Nope. I'd better explain."

"Go ahead."

"You know about Gödel?"

"I know Gödel's proof, but tell me from scratch. You might be doing some illegal reasoning."

He leaned forward as if he were going to tell me a dirty story.

"Well," he said, lowering his voice, "you're familiar with the conclusion that no machinelike entity that proceeds by clearly defined mechanical steps can complete any system that is rich enough to generate simple arithmetic—that is, make it a consistent system in which we could not come up with new, true, and still unproven propositions, in fact ones that would be unprovable in the system, yet clearly true."

"I know, math can't be mechanized."

"Not completely mechanized. We've done it to a remarkable degree . . ."

"What else is new?"

"Well, if Gödel's proof is true, and human minds can regularly generate true but unprovable propositions in any potentially self-consistent system, then mechanism, or determinism, does not apply to us."

"But what is it that you want to *do,* Witter?" I was only half listening. It was late in the day. The cafeteria was nearly empty and the newly polished floor was a large mirror; our booth seemed to float on it.

"Well," he said, "I want to give the new Artificial Intelligence the command to complete mathematics."

"What?" I suddenly saw what he was getting at.

"Don't you see? We can do an experiment that might settle the nature of the universe—whether we live in a hard determinism or a soft one in which free will is possible."

I smiled, feeling superior. "But we know Gödel was right. Math can't be completed. He gave a powerful formal proof, one in which you can't have it both ways."

Witter, who had been looking away as we spoke, turned his head half around and fixed me with one glassy brown eye. "Come on, Bruno. Why not run the experiment anyway?"

I shrugged and sat back, looking around. "As you said, it might take a long time—forever, if Gödel's right."

"Maybe," he said, finally looking at me with both eyes. The combination of the blue and brown eyes had always given me the creeps. "According to Gödel, the computer will crank out mathematical statements forever, and we'll never know if the body of the system is a complete one. But if it is complete, then our AI will finish it off in some finite period of time. It's the fastest system ever developed, able to do involved operations that might take centuries otherwise."

"No matter how fast it is, we won't disprove Gödel. He proved that, independently of all need to do experiments! Now I know why you want a lot of time. We won't live long enough to learn the result, even if you're right, which you can't be." I started to get up.

"Look," he insisted, "why not do the experiment? If we live in a hard determinism, as so many believe, then it's already true—the AI will complete math or any system we give it. But if Gödel is right, the AI-5 will run on forever, unable to complete."

"We don't have forever. You've gone bonkers."

"Why don't we do it? We *can* do the experiment! Look, for the first time an experiment involving pure logic and math may yield knowledge of the world outside."

That part appealed to me, but I saw a way of being perverse. Was he presenting me with a choice or dictating that I authorize the experiment?

He smiled, anticipating my thought. "Either there's free will, or you're fated to let the experiment be done."

I sighed. "But there's still the matter of how long it will take, Felix. AI-5, no matter how fast it is, may keep running and we won't be able to tell whether it's an uncompletable process or just a very long one."

He shrugged. "Aren't you willing to take the chance?"

"This just doesn't make any sense to me at all."

He smiled again. "But it gets to you, doesn't it? My point holds. Why not do it? Just to see. How often in the history of math or logic has there been a chance to do pure theoretical work that might reveal something about the real world?"

"But it's doomed to fail!"

He nodded. "Probably, Bruno, I'll grant you that. But even so, the experiment will be historic. Purely mathematical and empirical at the same time."

"Romantic mathematics, I call it."

"Or Kant's synthetic a priori!"

I'd read some of that metaphysical junk, and he seemed to be stretching it. Sure, synthetic meant acquiring new knowledge, and a priori meant that it wasn't derived from experience, strictly speaking, but from reasoning. Our experiment would give us new knowledge of the universe through nonempirical means. "But you're cheating," I said. "Whatever you call it, using the AI means only doing an empirical experiment."

He cocked one eyebrow and gave me a crazed stare with his blue eye. "Would you say that it would be more empirical if we did it by pencil and paper? That's all Gödel had to work with."

"Okay, I guess I'll have to say that there are no purely a priori activities. Even using the mind alone is a way of reaching out into the universe. What we call experiments are merely corroborations. Einstein himself said that if the experiments didn't come out as he expected, then he'd pity the God who made the universe that way."

"Okay, Bruno, I know you know more than most section chiefs, but are we going to do it or not?"

So we ran the experiment, if you could call it that. Witter was right about one thing. If Gödel's proof was somehow wrong, and we could complete even one system on our fast AI, then a lot of people would have to do a lot of rethinking in the groundwork of logic and math.

But I knew damn well that Gödel couldn't be wrong. Formal proofs do not fall easily. It would be a mistake of some kind if our AI-5 showed that completeness in a significant system was achievable.

All right. We both wanted to see what would happen if we tried it. We pieced the time together from a dozen other projects when people would be away or on vacation.

It was Friday night, after hours. We would be alone until Monday. I sat down at the keyboard and tapped in the command. Witter was sitting next to me, staring up at the bank of screens.

The AI began its run, building arithmetic up out of baby talk. Soon it was all going by in a blur, but the AI showed no sign of slowing down.

"There is one danger," I said as we sat back and waited. "If the AI can't complete arithmetic, it will sift through larger and larger banks of information . . ."

"It can handle infinite amounts of data," Witter replied.

"Yes, but the power needed for that, Witter, the power! The cost!"

He shook his head. "Don't shout. That won't happen. It will all be over in a few hours at most."

* * *

But the AI-5 kept running. An hour went by.

"It's not going to stop, Felix. It can't. Gödel was right. But even if he was wrong, it may take more than our lifetimes to prove it."

"Take it easy, Bruno. Go polish the floor or something." He was too serene.

Another hour went by. Witter stared at the screens, hypnotized by the blurred flow. Rivers of reasoning ran from their headwaters to a new ocean of well-formed propositions, and still the ocean was not filled; it would never be filled.

As I looked around at the clean right angles of the room, at the symmetrical terminals and easily accessed units, I began to think that maybe Witter was slightly stupid, that he didn't understand simple logic or the idea of a proof. Gödel's paradoxical conclusion could not be broken, unless it wasn't a double bind to begin with, because you can't have it both ways. Something was very wrong with Felix Witter.

And yet I wanted him to have a point. This was an experiment, a recourse to more than personal opinion; it could do more, in principle, than reasoning, prediction, or guesswork. Set a powerful genie to do the impossible—not because you think the genie can do it, but because you can ask, and it has the power to do all that's possible. So why not ask, just to see; human beings have always been suspicious of mere reasoning, no matter how powerful. Suddenly I wanted to see Gödel fall, to see the pride and arrogance of mathematicians crumble.

But as we watched the AI-5 chase the mirage, there was no sign of an end, no slowdown at all.

"I'm hungry," I said. "Want a pizza?"

He nodded without looking at me. I got up, went out into the hall, and called it in from the wall phone. Then I alerted the security guard downstairs and asked him to leave it out on a cart in front of our workroom.

"We may have to stop it," I said hours later, "even if it's close to completion." Though the pizza had been very bad, I thought as I eyed the empty boxes on the cart, a full stomach had taken some of the romance out of what we were doing. "We can't tie up all this power and time indefinitely. It's using more every minute, and it'll be my ass if we can't justify it."

"No!" Witter shouted maniacally. "It may be very close."

I burped, waiting for my heartburn to subside. The AI-5 hummed along.

"We can continue from this point onward at another time," I insisted.

"Be quiet!"

I reached over to stop the run. Felix grabbed my hand and pinned it to the panel.

"What's wrong with you?" I demanded.

"Just a few minutes more," he said, fixing me with his mismatched eyes. "We're at the edge of a major discovery!"

"Felix, this can't be done." I struggled to free myself, but his strength was that of a true believer.

"Be still, you fool," he said harshly. "Don't you see? This will be the culmination of our careers. We'll never match this no matter how hard we work. Gödel is one of the supreme monuments of mathematics, marking the limits of human minds. If we topple him . . ."

"You may not like what you get," I said, twisting my arm. "If his proof is right, then mechanism is false and minds are not machines. They escape the completeness of the purely mechanical. But if Gödel is wrong, then we're automatons! I'd rather not know."

He shook his head. "There's even more to it than that, Bruno."

"What?" I was breathing very hard, unable to free myself.

"We're opening up the very vitals of reality."

I had to laugh. "By manipulating man-made symbolic structures? You need a bucket of cold water to soak your head in. Let me go!"

"Completion may be only a few minutes away. Do you want to stop and then wonder what might have been?" He tightened his grip.

"But you can't know how far along it is."

He let go of my hand and seemed to cool down, and I found I didn't have the heart to reach over and stop the run.

"You're right," he said, "I'm sorry. It probably is all for nothing."

I massaged my hand. The AI continued its work run. "Don't feel too bad about it," I managed to say. "It was a nice idea, but it had to confirm Gödel. I'm glad we're not machines."

He was shaking his head. "You don't understand. There's no reason to fear that. It's not a problem."

"What isn't?"

"Free will," he said as the AI-5 stopped its run.

Witter and I looked at each other, then at the main screen. It read:

SYSTEM CAPABLE OF GENERATING
ARITHMETIC COMPLETE

"It's a mistake of some kind," I said. Something strange seemed to pass across my eyes. I sat back, expecting to lose consciousness as the tension got to me.

"Maybe," Witter was saying, "but we can test to see if it's a mistake."

"How?" I heard myself ask, even though I knew the answer.

"By trying to make a true statement that is not provable in the system. As long as the AI can show us that we can't make such a statement by proving it, then the system is complete."

The room went black for a second. "But maybe we can't make such a statement," I said.

"We can try," he answered.

We tried for the next twelve hours. I was relieved that our prime AI was no longer running a huge power draw. Witter brought a smaller AI on-line and had it question the alleged complete system achieved by the AI-5. It failed to come up with a single true proposition that was not provable in the complete system.

"There's no question about it," Felix said finally.

"There's only one thing left to do," I replied. "We've got to run the whole thing again."

Witter looked at me, smiled strangely, then sat down and gave the command.

As the AI-5 began its second run at Gödel, Witter turned to me and said, "Funny about determinism. I always think of it as stuff outside me, pushing at my skin. But I feel free inside. When that second run finishes, we'll be certain that we're living in a hard determinism. No choice is our own, if we've understood the word correctly. Even our decision to run the AI-5 again was not made freely. We're automatons. No avoiding the conclusion, Bruno."

He was baiting me, I was sure. "But we resist the notion. Doesn't that suggest something?"

He shrugged. "That we're free in our minds but not in our actions. We can envision alternatives, but whichever one we pick is determined, right up through an infinite future."

"Witter, I thought you were intelligent. There can't be such a thing as unconditioned freedom. There are always initial conditions—necessary and sufficient conditions for every choice. Otherwise we could perform miracles, make happen things that are uncaused. The existence of free will cannot violate causality."

He grimaced at me and I felt stupid. "Yeah, I know all that. But *do* we have the freedom to choose between alternatives?"

"I think we do. Physical conditions make us both the determined and determinators in our own right. Things affect us and we affect them. Determinism goes right down into us, into our consciousness and will, and we send it back out. I couldn't prove it to you without a physiologist, though."

The AI-5 was still running its second completion smoothly. If it succeeded, then it might be that we were living in a universe where even choice among alternatives was an illusion.

Witter looked at me suddenly. "I wonder if our running this program can have an effect on the universe we live in?"

"What are you talking about?" I asked. He seemed to have a mind like a break dancer.

"Maybe our attempting what Gödel said was impossible can change the universe?"

"I don't think so, Felix. But there are other things you might like to consider."

He took a deep breath. "What's that?"

"Well, we began with the idea that no finitary deductive system can complete a rich, self-consistent system. But what if the AI-5 is not a finitary deductive system? Assume it can work outside the limits of the human mind, which is all that Gödel may have charted. It was all he could demonstrate because he had only his own mind to work with."

Witter nodded. "I see what you mean. If our AI reaches completion, then it follows, perhaps, that it's not a finitary deductive system, and we can draw no conclusions about the nature of the universe."

I smiled. "Right. And we don't have to worry about being automatons, or that our sense of inner freedom is a mirror trick of some kind. Free will is a special case of determinism. It's determinism from the inside. The means of determinism are also those of free will."

Witter was watching the screen with a worried look on his face, as if he now expected the AI to fail. It didn't matter one way or the other, if what I had said was true.

"Unconditioned free will would be omnipotence," I continued, "and that's an absurd state to be in. No law, no causal structures. It's just a conceptual extreme, like infinities."

"Something is working against us," Witter said softly.

"What do you mean?"

He gripped the panel. "It won't come out the same way twice," he replied.

"You're still mistaking the maps for reality," I said.

"Look at the time, you fool! It's almost as long as before. If the AI doesn't repeat its completion in the same time, it will run on forever."

"So what. We have the first completion in memory, step by step, for whatever it's worth."

He swiveled his chair and glared at me. His eyes were bloodshot and had dark circles around them. The whole experiment, I saw, was eating up his entire energy. "You don't see, do you?" he said. "You think in terms of tricks of language, ways of speaking . . . you can't imagine worlds dying and others supplanting them. You don't give a damn about anything except apportioning time and keeping other administrators happy."

"What are you talking about, Felix? I'm here with you, and we're doing what you wanted. Have you lost your mind?" I almost felt hurt, as if he were questioning my loyalty.

He pointed to the clock on the wall. "Look, time's up and our AI is still running."

"So what? It was a fluke the first time, a mistake. You can't beat Gödel, and it wouldn't matter if you could."

He laughed. "You still don't see!"

"No, I don't."

The AI-5 was still running.

"It will run forever this time. Our decision to run the experiment puts us at a great juncture between possible universes. We collapsed the wave function reaching our minds."

"What are you saying?" I demanded.

"Proving that *our* universe was deterministic threw us into a freer one. Gödel proved his work in the wrong universe. Here the AI will run forever. But if we stop it and start again, something even stranger might happen."

"You're off the deep end now," I said, feeling sorry for him.

"We might be moving across a whole series of universes, drawing closer to the unconditional omnipotence that has the true freedom to be everything . . ."

"Yeah, and can't become anything in particular. That's what I was saying. Witter, wake up. We have the other program. Go see for yourself. That system was completed. In this one there's obviously some kind of difficulty. Neither result means a thing. Get that through your stupid head!" Mathematicians were all idealists to some degree or another, always secretly believing in the literal existence of infinities, numbers, and tortured geometries. Witter was no exception.

He shook his head and smiled. "There's nothing in the memory, Bruno. See for yourself. Go ahead, punch it up."

I leaned forward and punched in the order. Nothing came up. I went into search mode. Still nothing.

"We've left that universe behind," Witter said.

"It's got to be here," I said.

The screen remained blank.

"You erased the memory!" I shouted.

"I did not," he replied softly, and I knew he was telling the truth.

I glanced at the food cart; it winked out of existence.

"Did you see that?" I asked.

"Bruno!" Witter shouted. "We've escaped a totalitarian cosmos. We're free!"

"Relatively," I said, shaken.

He was looking at me strangely, and I saw that both his eyes were now brown. As the AI-5 continued its endless run into a free infinity, I feared what we would find when we went outside . . .

DOUGLAS HOFSTADTER

The Tale of Happiton

Happiton was a happy little town. It had 20,000 inhabitants, give or take seven, and they were productive citizens who mowed their lawns quite regularly. Folks in Happiton were pretty healthy. They had a life expectancy of seventy-five years or so, and lots of them lived to ripe old ages. Down at the town square, there was a nice big courthouse with all sorts of relics from WW II and monuments to various heroes and whatnot. People were proud, and had the right to be proud, of Happiton.

On the top of the courthouse, there was a big bell that boomed every hour on the hour, and you could hear it far and wide—even as far out as Shady Oaks Drive, way out nearly in the countryside.

One day at noon, a few people standing near the courthouse noticed that right after the noon bell rang, there was a funny little sound coming from up in the belfry. And for the next few days, folks noticed that this scratching sound was occurring after every hour. So on Wednesday, Curt Dempster climbed up into the belfry and took a look. To his surprise, he found a crazy kind of contraption rigged up to the bell. There was this mechanical hand, sort of a robot arm, and next to it were five weird-looking dice that it could throw into a little pan. They all had twenty sides on them, but instead of being numbered one through twenty, they were just numbered zero through nine, but with each digit appearing on two opposite sides. There was also a TV camera that pointed at the pan and it seemed to be attached to a microcomputer or something. That's all Curt could figure out. But then he noticed that on top of the computer, there was a neat little envelope marked "To the friendly folks of Happiton." Curt decided that he'd take it downstairs and open it in the presence of his friend the mayor, Janice Fleener. He found Janice easily enough, told her about what he'd found, and then they opened the envelope. How neatly it was written! It said this:

Grotto 19, Hades
June 20, 1983

Dear folks of Happiton,

I've got some bad news and some good news for you. The bad first. You know your bell that rings every hour on the hour? Well, I've set it up so that each time it rings, there is exactly one chance in a hundred thousand—that is, 1/100,000—that a Very Bad Thing will occur. The way I determine if that Bad Thing will occur is, I have this robot arm fling its five dice and see if they all land with "7" on top. Most of the time, they won't. But if they do—and the odds are exactly 1 in 100,000—then great clouds of an unimaginably revolting-smelling yellow-green gas called "Retchgoo" will come oozing up from a dense network of underground pipes that I've recently installed underneath Happiton, and everyone will die an awful, writhing, agonizing death. Well, that's the bad news.

Now the good news! You all can prevent the Bad Thing from happening, if you send me a bunch of postcards. You see, I happen to like postcards a whole lot (especially postcards of Happiton), but to tell the truth, it doesn't really much matter what they're of. I just *love* postcards! Thing is, they have to be written personally—not typed, and especially not computer-printed or anything phony like that. The more cards, the better. So how about sending me some postcards—batches, bunches, boxes of them?

Here's the deal. I reckon a typical postcard takes you about 4 minutes to write. Now suppose just one person in all of Happiton spends 4 minutes one day writing me, so the next day, I get one postcard. Well, then, I'll do you all a favor: I'll slow the courthouse clock down a bit, for a day. (I realize this is an inconvenience, since a lot of you tell time by the clock, but believe me, it's a lot more inconvenient to die an agonizing, writhing death from the evil-smelling, yellow-green Retchgoo.) As I was saying, I'll slow the clock down for *one day,* and by how much? By a factor of 1.00001. Okay, I know that doesn't sound too exciting, but just think if all 20,000 of you send me a card! For *each* card I get that day, I'll toss in a slow-up factor of 1.00001, the next day. That means that by sending me 20,000 postcards a day, you all, working together, can get the clock to slow down by a factor of *1.00001 to the 20,000th power,* which is just a shade over 1.2, meaning it will ring every 72 minutes.

All right, I hear you saying, "72 minutes is just barely over an hour!" So I offer you more! Say that one day I get 160,000 postcards (heavenly!). Well then, the very next day I'll show my gratitude by slowing your clock down, all day long, midnight to midnight, by 1.00001 to the 160,000th power, and that ain't chickenfeed. In fact, it's about 5, and that means the clock will ring only every 5 hours, meaning those sinister dice will only get rolled about 5 times (instead of the usual 24). Obviously, it's better for both of us that way. You have to bear in mind that I don't have any personal interest in seeing that awful Retchgoo come rushing and gushing up out of those pipes and causing every last one of you to perish in grotesque, mouth-foaming, twitching convulsions. All I care about is getting postcards! And to send me 160,000 a day wouldn't cost you folks that much effort, being that it's just 8 postcards a day—just about a half hour a day for each of you, the way I reckon it.

So my deal is pretty simple. On any given day, I'll make the clock go off once every X hours, where X is given by this simple formula:

$$X = 1.00001^N$$

Here, N is the number of postcards I received the previous day. If N is 20,000, then X will be 1.2, so the bell would ring 20 times per day, instead of 24. If N is 160,000, then X jumps way up to about 5, so the clock would slow way down—just under 5 rings per day. If I get *no* postcards, then the clock will ring once an hour, just as it does now. The formula reflects that, since if N is 0, X will be 1. You can work out other figures yourself. Just think how much safer and securer you'd all feel knowing that your courthouse clock was ticking away so slowly!

I'm looking forward with great enthusiasm to hearing from you all.

Sincerely yours,
Demon #3127

The letter was signed with beautiful medieval-looking flourishes, in an unusual shade of deep red . . . ink?

"Bunch of hogwash!" spluttered Curt. "Let's go up there and chuck the whole mess down onto the street and see how far it bounces." While he was saying this, Janice noticed that there was a smaller note clipped onto the back of the last sheet, and turned it over to read it. It said this:

> *P.S.*—It's really not advisable to try to dismantle my little setup up there in the belfry: I've got a hair trigger linked to the gas pipes, and if anyone tries to dismantle it, pssssst! Sorry.

Janice Fleener and Curt Dempster could hardly believe their eyes. What gall! They got straight on the phone to the police department, and talked to Officer Curran. He sounded poppin' mad when they told him what they'd found, and said he'd do something about it right quick. So he hightailed it over to the courthouse and ran up those stairs two at a time, and when he reached the top, a-huffin' and a-puffin', he swung open the belfry door and took a look. To tell the truth, he was a bit ginger in his inspection, because one thing Officer Curran had learned in his many years of police experience is that an ounce of prevention is worth a pound of cure. So he cautiously looked over the strange contraption, and then he turned around and quite carefully shut the door behind him and went down. He called up the town sewer department and asked them if they could check out whether there was anything funny going on with the pipes underground.

Well, the long and the short of it is that they verified everything in the Demon's letter, and by the time they had done so, the clock had struck five more times and those five dice had rolled five more times. Janice Fleener had in fact had her thirteen-year-old daughter Samantha go up and sit in a wicker chair right next to the microcomputer and watch the robot arm throw those dice. According to Samantha, an occasional seven had turned up now and then, but never had two sevens shown up together, let alone sevens on all five of the weird-looking dice!

The next day, the *Happiton Eagle-Telephone* came out with a front-page story telling all about the peculiar goings-on. This caused quite a commotion. People *everywhere* were talking about it, from Lidden's Burger Stop to Bixbee's Druggery. It was truly the talk of the town.

When Doc Hazelthorn, the best pediatrician this side of the Cornyawl River, walked into Ernie's Barbershop, corner of Cherry and Second, the atmosphere was more somber than usual. "Whatcha gonna do, Doc?" said Big Ernie, the jovial barber, as he was clipping the few remaining hairs on old Doc's pate. Doc (who was also head of the Happiton City Council) said the news had come as quite a shock to him and his family. Red Dulkins, sitting in the next chair over from Doc, said he felt the same way. And then the two gentlemen waiting to get their hair cut both added their words of agreement. Ernie, summing it up, said the whole town seemed quite upset. As Ernie removed the white smock from Doc's lap and shook the hairs off it, Doc said that he had just decided to bring the matter up first thing at the next City Council meeting, Tuesday evening. "Sounds like a good idea, Doc!" said Ernie. Then Doc told Ernie he couldn't make the usual golf date this weekend because some friends of his had invited him to go fishing out at Lazy Lake, and Doc just couldn't resist.

Two days after the Demon's note, the *Eagle-Telephone* ran a feature article in which many residents of Happiton, some prominent, some not so prominent, voiced their opinions. For instance, eleven-year-old Wally Thurston said he'd gone out and bought up the whole supply of picture postcards at the 88-Cent Store, $14.22 worth of postcards, and he'd already started writing a few. Andrea McKenzie, sophomore at Happiton High, said she was really worried and had had nightmares about the gas, but her parents told her not to worry, things had a way of working out. Andrea said maybe her parents weren't taking it so seriously because they were a generation older and didn't have as long to look forward to anyway. She said she was spending an hour each day writing postcards. That came to fifteen or sixteen cards each day. Hank Hoople, a janitor at Happiton High, sounded rather glum: "It's all fate. If the bullet has your name on it, it's going to happen, whether you like it or not." Many other citizens voiced concern and even alarm about the recent developments.

But some voiced rather different feelings. Ned Furdy, who as far as anyone could tell didn't do much other than hang around Simpson's bar all day (and most of the night) and buttonhole anyone he could, said, "Yeah, it's a problem, all right, but I don't know nothin' about gas and statistics and such. It should all be left to the mayor and the Town Council, to take care of. They know what they're doin'. Meanwhile, eat, drink, and be merry!" And Lulu Smyth, seventy-seven-year-old proprietor of Lulu's Thread 'N' Needles Shop, said "I think it's all a ruckus in a teapot, in my opinion. Far as I'm concerned, I'm gonna keep on sellin' thread 'n' needles, and playin' gin rummy every third Wednesday."

When Doc Hazelthorn came back from his fishing weekend at Lazy Lake, he had some surprising news to report. "Seems there's a demon left a similar setup in the church steeple down in Dwaynesville," he said. (Dwaynesville was the next town down the road, and the arch-rival of Happiton High in football.) "The Dwaynesville demon isn't threatening them with gas, but with radioactive water. Takes a little longer to die, but it's just as bad. And I hear tell there's a demon with a subterranean volcano up at New Athens." (New Athens was the larger town twenty miles up the Cornyawl from Dwaynesville, and the regional center of commerce.)

A lot of people were clearly quite alarmed by all this, and there was plenty of arguing on the streets about how it had all happened without anyone knowing. One thing that was pretty universally agreed on was that a commission should be set up as soon as possible, charged from here on out with keeping close tabs on all subterranean activity within the city limits, so that this sort of outrage could never happen again. It appeared probable that Curt Dempster, who was the moving force behind this idea, would be appointed its first head.

Ed Thurston (Wally's father) proposed to the Jaycees (of which he was a member in good standing) that they donate $1,000 to support a postcard-writing campaign by town kids. But Enoch Swale, owner of Swale's Pharmacy and the Sleepgood Motel, protested. He had never liked Ed much, and said Ed was proposing it simply because his son would gain status that way. (It was true that Wally had recruited a few kids and that they spent an hour each afternoon after school writing cards. There had been a small article in the paper about it once.) After considerable debate, Ed's motion was narrowly defeated. Enoch had a lot of friends on the City Council.

Nellie Doobar, the math teacher at High, was about the only one who checked out the Demon's math. "Seems right to me," she said to the reporter who called her about it. But this set her to thinking about a few things. In an hour or two, she called back the paper and said, "I figured something out. Right now, the clock is still ringing very close to once every hour. Now there are 720 hours per month, and so that means there are 720 chances each month for the gas to get out. Since each chance is 1 in 100,000, it turns out that each month, there's a bit less than a 1-in-100 chance that Happiton will get gassed. At that rate, there's about eleven chances in twelve that Happiton will make it through each year. That may sound pretty good, but the chances we'll make it through any eight-year period are almost exactly fifty-fifty, exactly the same as tossing a coin. So we can't really count on very many years . . ."

This made big headlines in the next afternoon's *Eagle-Telephone*—in fact, even bigger than the plans for the County Fair! Some folks started calling up Mrs. Doobar anonymously and telling her she'd better watch out what she was saying if she didn't want to wind up with a puffy face or a fat lip. Seems like they couldn't quite keep it straight that Mrs. Doobar wasn't the one who'd set the thing up in the first place.

After a few days, though, the nasty calls died down pretty much. Then Mrs. Doobar called up the paper again and told the reporter, "I've been calculating a bit more here, and I've come up with the following, and they're facts every last one of them. If all 20,000 of us were to spend half an hour a day writing postcards to the Demon, that would amount to 160,000 postcards a day, and just as the Demon said, the bell would ring pretty near every *five* hours instead of every hour, and that would mean that the chances of us getting wiped out each month would go down considerable. In fact, there would only be about one chance in 700 that we'd go down the tubes in any given month, and only about a chance in 60 that we'd get zapped each year. Now I'd say that's a darn sight better than one chance in twelve per year, which is what it is if we don't write any postcards (as is more or less the case now, except for Wally Thurston and Andrea McKenzie and a few other kids I heard of). And for every eight-year period, we'd only be running a 13 percent risk instead of a 50 percent risk."

"That sounds pretty good," said the reporter cheerfully.

"Well," replied Mrs. Doobar, "it's not too bad, but we can get a whole lot better by doublin' the number of postcards."

"How's that, Mrs. Doobar?" asked the reporter. "Wouldn't it just get twice as good?"

"No, you see, it's an exponential curve," said Mrs. Doobar, "which means that if you *double N*, you *square X*."

"That's Greek to me," quipped the reporter.

"*N* is the number of postcards and *X* is the time between rings," she replied quite patiently. "If we all write a half hour a day, *X* is five hours. But that means that if we all write a whole *hour* a day, like Andrea McKenzie in my algebra class, *X* jumps up to twenty-five hours, meaning that the clock would ring only about once a day, and obviously that would reduce the danger a *lot*. Chances are, hundreds of years would pass before five sevens would turn up together on those infernal dice. Seems to me that under those circumstances, we could pretty much live our lives without worrying about the gas at all. And that's for writing about an hour a day, each one of us."

The reporter wanted some more figures detailing how much different amounts of postcard writing by the populace would pay off, so Mrs. Doobar obliged by going back and doing some more figuring. She figured out that if 10,000 people—half the population of Happiton—did two hours a day for the year, they could get the same result—one ring every twenty-five hours. If only 5,000 people spent two hours a day, or if 10,000 people spent one hour a day, then it would go back to one ring every five hours (still a lot safer than one every hour). Or, still another way of looking at it, if just 1,250 of them worked *full-time* eight hours a day), they could achieve the same thing.

"What about if we all pitch in and do four minutes a day, Mrs. Doobar?" asked the reporter.

"Fact is, 'twouldn't be worth a damn thing! (Pardon my French.)" she replied. "*N* is 20,000 that way, and even though that sounds pretty big, *X* works out to be just 1.2, meaning one ring every 1.2 hours, or 72 minutes. That way, we still have about a chance of 1 in 166 every month of getting wiped out, and 1 in 14 every year of getting it. Now that's real scary, in my book. Writing cards only starts making a noticeable difference at about fifteen minutes a day per person."

By this time, several weeks had passed, and summer was getting into full swing. The County Fair was buzzing with activity, and each evening after folks came home, they could see loads of fireflies flickering around the trees in their yards. Evenings were peaceful and relaxed. Doc Hazelthorn was playing golf every weekend, and his scores were getting down into the low nineties. He was feeling pretty good. Once in a while he remembered the Demon, especially when he walked downtown and passed the courthouse tower, and every so often he would shudder. But he wasn't sure what he and the City Council could do about it.

The Demon and the gas still made for interesting talk, but were no longer such big news. Mrs. Doobar's latest revelations made the paper, but were relegated this time to the second section, two pages before the comics, right next to the daily horoscope column. Andrea McKenzie read the article avidly and showed it to a lot of her school friends, but to her surprise it didn't seem to stir up much interest in them. At first, her best friend, Kathi Hamilton, a very bright girl who had plans to go to State and major in history, enthusiastically joined Andrea and wrote quite a few cards each day. But after a few days Kathi's enthusiasm began to wane.

"What's the point, Andrea?" Kathi asked. "A handful of postcards from me isn't going to make the slightest bit of difference. Didn't you read Mrs. Doobar's article? There have got to be 160,000 a day to make a big difference."

"That's just the point, Kath!" replied Andrea exasperatedly. "If you and everyone else will just do your part, we'll *reach* that number—but you can't cop out!" Kathi didn't see the logic, and spent most of her time doing her homework for the summer school course in world history she was taking. After all, how could she get into State if she flunked world history?

Andrea just couldn't figure out how come Kathi, of all people, so interested in history and the flow of time and world events, could not see her *own* life being touched by such factors, so she asked Kathi, "How do you know there will be any *you* left to *go* to State, if you don't write postcards? Each year, there's a one-in-twelve chance of you and me and all of us being wiped out! Don't you even want to work against that? If people would just *care,* they could *change* things! An hour a day! Half an hour a day! Fifteen minutes a day!"

"Oh, come *on,* Andrea!" said Kathi annoyedly. "Be realistic."

"Darn it all, *I'm* the one who's *being* realistic," said Andrea. "If you don't help out, you're adding to the burden of someone else."

"For Pete's sake, Andrea," Kathi protested angrily, "I'm *not* adding to anyone else's burden. Everyone can help out as much as they want, and no one's obliged to do anything at all. Sure, I'd like it if everyone were helping, but you can see for yourself, practically nobody is. So I'm not going to waste my time. I need to pass world history."

And sure enough, Andrea had to do no more than listen each hour, right on the hour, to hear that bell ring to realize that nobody was doing much. It once had sounded so pleasant and reassuring, and now it sounded creepy and ominous to her, just like the fireflies and the barbecues. Those fireflies and barbecues really bugged Andrea, because they seemed so *normal*, so much like any *other* summer—only *this* summer was *not* like any other summer. Yet nobody seemed to realize that. Or rather, there was an undercurrent that things were not quite as they should be, but nothing was being done . . .

One Saturday, Mr. Hobbs, the electrician, came around to fix a broken refrigerator at the McKenzies' house. Andrea talked to him about writing postcards to the Demon. Mr. Hobbs said to her, "No time, no time! Too busy fixin' air conditioners! In this heat wave, they been breakin' down all over town. I work a ten-hour day as it is, and now it's up to eleven, twelve hours a day, includin' weekends. I got no time for postcards, Andrea." And Andrea saw that for Mr. Hobbs it was true. He had a big family and his children went to parochial school, and he had to pay for them all, and . . .

Andrea's older sister's boyfriend, Wayne, was a star halfback at Happiton High. One evening he was over and teased Andrea about her postcards. She asked him, "Why don't you write any, Wayne?"

"I'm out lifeguardin' every day, and the rest of the time I got scrimmages for the fall season."

"But you could take some time out—just fifteen minutes a day—and write a few postcards!" she argued. He just laughed and looked a little fidgety. "I don't know, Andrea," he said. "Anyway, me 'n' Ellen have got better things to do—huh, Ellen?" Ellen giggled and blushed a little. Then they ran out of the house and jumped into Wayne's sports car to go bowling at the Happi-Bowl.

* * *

Andrea was puzzled by all her friends' attitudes. She couldn't understand why everyone had started out so concerned but then their concern had fizzled, as if the problem had gone away. One day when she was walking home from school, she saw old Granny Sparks out watering her garden. Granny, as everyone called her, lived kitty-corner from the McKenzies and was always chatty, so Andrea stopped and asked Granny Sparks what she thought of all this. "Pshaw! Fiddlesticks!" said Granny indignantly. "Now Andrea, don't you go around believin' all that malarkey they print in the newspapers! Things are the same here as they always been. I oughta know—I've been livin' here nigh on eighty-five years!"

Indeed, that was what bothered Andrea. *Everything* seemed so annoyingly ***normal***. The teenagers with their cruising cars and loud motorcycles. The usual boring horror movies at the Key Theater down on the square across from the courthouse. The band in the park. The parades. And especially, the damn fireflies! Practically nobody seemed moved or affected by what to her seemed the most overwhelming news she'd ever heard. The only other truly sane person she could think of was little Wally Thurston, that eleven-year-old from across town. What a ridiculous irony, that an eleven-year-old was saner than all the adults!

Long about August 1, there was an editorial in the paper that gave Andrea a real lift. It came from out of the blue. It was written by the paper's chief editor, "Buttons" Brown. He was an old-time journalist from St. Joe. Missouri. His editorial was real short. It went like this:

The Disobedi-Ant

The story of the Disobedi-Ant is very short. It refused to believe that its powerful impulses to play instead of work were anything but unique expressions of its very unique self, and it went its merry way, singing, "What I choose to do has nothing to do with what any-ant else chooses to do! What could be more self-evident?"

Coincidentally enough, so went the reasoning of all its colonymates. In fact, the same refrain was independently invented by every last ant in the colony, and each ant thought it original. It echoed throughout the colony, even with the same melody.

The colony perished.

Andrea thought this was a terrific allegory, and showed it to all her friends. They mostly liked it, but to her surprise not one of them started writing postcards.

All in all, folks were pretty much back to daily life. After all, nothing much seemed really to have changed. The weather had turned real hot, and folks congregated around the various swimming pools in town. There were lots of barbecues in the evenings, and every once in a while somebody'd make a joke or two about the Demon and the postcards. Folks would chuckle and then change the topic. Mostly, people spent their time doing what they'd always done, and enjoying the blue skies. And mowing their lawns regularly, since they wanted the town to look nice.

DON SAKERS

The Finagle Fiasco

Yes, I remember the Murphy episode. Of course, I was not Grand Master of Euler at the time—I was only Assistant Christensen Professor of Topology. Still, I don't suppose anyone will ever forget that time, when the Math Institute here on Euler was all that stood between the galaxy and total domination by a sadistic megalomaniac.

What's that? Oh, yes, I know the Psychology Institute has done penance for allowing Khar-Davii to take over. And I understand that they say it can never happen again. Well, I wonder—psychology is not of course an exact discipline, like math.

Eh? Yes, the Murphy episode. As I recall, it was shortly after the spring term had begun. I had trouble with some of my displays; the Twenty-Dimension Simulator had developed a singularity, and simply would not accept fields with more than eighty operations. Maintenance told me that the entire system would have to be shut down for reprogramming, and I went to the Grand Master for approval. She was conversing on the hyperwave; I waited until she was done. In due time she opened the privacy hood and smiled at me. "Ah, Professor Yagwn. How are you?"

"Fine, Madam. And yourself?"

She sighed. "I could be better, Yagwn. You've heard of this Khar-Davii, who calls himself the Conqueror? Well, it appears that he has taken over the Galactic Council and killed the Co-ordinator. He has proclaimed himself Monarch of Humanity, and the inhabited worlds are falling all over their own feet to surrender to him."

I recalled hearing something about the matter on the news. "Are his weapons that formidable?"

"Apparently so. Euler is the only planet that has not yielded. I was just talking with the outlying Galactic Traffic station—Khar-Davii's fleet is even now heading toward this world." She glanced at a data screen on her desk. "Ah, excuse me. The fleet has arrived. We are surrounded."

I had no opportunity to voice an opinion. There was a bright flash of light, and suddenly the image of a corpulent human man appeared in the center of the Grand Master's office. Behind him were banks of machinery tended by warriors in full battle dress.

"I am Khar-Davii, the Conqueror. Your miserable planet has refused to accept my rule. You will surrender to me now or I will destroy your world."

I suppressed a grin; the Grand Master did not bother to hide her amusement. "I hardly think it is a miserable world. I rather like it. Conqueror, your plan of conquest would interfere with our spring term, and I'm sure that the commotion would upset many of our scholars."

Khar-Davii narrowed his brows. "As I was told—you are totally out of touch with the real universe. Mathematicians and philosophers—not a practical being in the bunch."

The Grand Master lost her smile. One thing that always bothered her was the accusation that Euler was out of touch with reality. To her, math was the highest form of reality. She stood and faced Khar-Davii.

"My dear Conqueror, I will not allow you to bother Euler. If you wish to attack, then do so—but let me show you something of our defenses first." She touched a button, and a screen behind her showed the image of a great cannon.

I drew in my breath sharply at the sight.

"And what is that machine, Grand Master?" Khar-Davii asked with a smile. "Will it shoot strings of numbers at us?"

The Grand Master answered with another smile. "No doubt, Conqueror, a man with your military background has heard of the Murphy laws? That which can go wrong, *will* go wrong. Here we have them formulated as a theorem, and implemented as a weapon."

"And this is your defense?"

She spread her hands and regarded him as though he were a simpleton—which seemed readily apparent. "Long ago we investigated the Murphy laws completely. This machine amplifies their effects. If you attack us, your guns will fail to fire, your ships will suffer instrument breakdowns, your most trusted officers will trip and accidentally sound recall orders. You could never beat us."

Khar-Davii dissolved in a fit of laughter. "My fleet has been listening to this conversation—now they know what 'terrible weapon' Euler will use against them." He stopped chuckling. "Grand Master, prepare for your death. Fleet—Attack!"

The attack did not last long.

Since I had a little time to spare, I watched it on the viewscreens from the Grand Master's office. After twenty minutes or so, only the Conqueror's flagship was left in fighting condition. It was not too long afterward that Khar-Davii's image reappeared in the office. The Conqueror was harried and bedraggled, and there was fear in his eyes.

"Can I help you, Conqueror?" the Grand Master asked.

"Enough. Enough. Turn off that machine. We will sue for peace. I will not attack your planet any more."

"Fine." She pressed another button. "Your treaty has been logged. We have other weapons that we can use against you, should you try to break your word. I will thank you now to take the remnants of your fleet away without bothering us . . . we have important work to do."

"You will not try to prevent me from ruling the Galaxy? Your Murphy Machine is a more formidable weapon than any I possess."

She smiled. "Poor, poor Conqueror. You should have taken more mathematics classes. Deductive reasoning would have helped you. The Murphy Machine worked perfectly—as soon as it was turned on, things started to go wrong. The first problem that developed was the failure of the machine itself."

"Failure. . . ?"

"Yes." She laughed. "It was the superstition of your crews that defeated you, Khar-Davii. They believed that they could not win, and so they did not."

Khar-Davii snarled, and his image vanished. Viewscreens showed his ships limping away from Euler.

"We shall have no more trouble from him. The memory of his defeat and his fear of a recurrence will prevent him from returning. He will attempt to rule the galaxy and will forget about Euler." She shook her head. "What can I do for you, Yagwn?"

"I need permission to shut down the Twenty-Dimension Simulator for reprogramming."

"Very well, I will make the necessary notifications."

"Thank you." I turned to go, then paused at the door. "About the Murphy Machine, Grand Master. Do you think it was kind to lie to him so?"

"Kinder than letting him know what a terrible power he is really up against. He thinks the machine unworkable." She shrugged. "Let *him* figure out why his empire dissolves so quickly."

Dismissing me, she bent her head back to her work.

LARRY NIVEN

Convergent Series

It was a girl in my anthropology class who got me interested in magic. Her name was Ann, and she called herself a white witch, though I never saw her work an effective spell. She lost interest in me and married somebody, at which point I lost interest in her; but by that time magic had become the subject of my thesis in anthropology. I couldn't quit, and wouldn't if I could. Magic fascinated me.

The thesis was due in a month. I had a hundred pages of notes on primitive, medieval, Oriental, and modern magic. Modern magic meaning psionics devices and such. Did you know that certain African tribes don't believe in natural death? To them, *every* death is due to witchcraft, and in every case the witch must be found and killed. Some of these tribes are actually dying out due to the number of witchcraft trials and executions. Medieval Europe was just as bad in many ways, but they stopped in time . . . I'd tried several ways of conjuring Christian and other demons, purely in a spirit of research, and I'd put a Taoist curse on Professor Pauling. It hadn't worked. Mrs. Miller was letting me use the apartment-house basement for experiments.

Notes I had, but somehow the thesis wouldn't move. I knew why. For all I'd learned, I had nothing original to say about anything. It wouldn't have stopped everyone (remember the guy who counted every *I* in *Robinson Crusoe*?) but it stopped me. Until one Thursday night—

I get the damnedest ideas in bars. This one was a beaut. The bartender got my untouched drink as a tip. I went straight home and typed for four solid hours. It was ten minutes to twelve when I quit, but I now had a complete outline for my thesis, based on a genuinely new idea in Christian witchcraft. All I'd needed was a hook to hang my knowledge on. I stood up and stretched . . .

. . . And knew I'd have to try it out.

All my equipment was in Mrs. Miller's basement, most of it already set up. I'd left a pentagram on the floor two nights ago. I erased that with a wet rag, a former washcloth, wrapped around a wooden block. Robes, special candles, lists of spells, new pentagram . . . I worked quietly so as not to wake anyone. Mrs. Miller was sympathetic; her sense of humor was such that they'd have burned her three centuries ago. But the other residents needed their sleep. I started the incantations exactly at midnight.

At fourteen past I got the shock of my young life. Suddenly there was a demon spread-eagled in the pentagram, with his hands and feet and head occupying all five points of the figure.

I turned and ran.

He roared, "Come back here!"

I stopped halfway up the stairs, turned, and came back down. To leave a demon trapped in the basement of Mrs. Miller's apartment house was out of the question. With that amplified basso profundo voice he'd have wakened the whole block.

He watched me come slowly down the stairs. Except for the horns he might have been a nude middle-aged man, shaved, and painted bright red. But if he'd been human you wouldn't have wanted to know him. He seemed built for all of the Seven Deadly Sins. Avaricious green eyes. Enormous gluttonous tank of a belly. Muscles soft and drooping from sloth. A dissipated face that seemed permanently angry. Lecherous—never mind. His horns were small and sharp and polished to a glow.

He waited until I reached bottom. "That's better. Now what kept you? It's been a good century since anyone called up a demon."

"They've forgotten how," I told him. "Nowadays everyone thinks you're supposed to draw the pentagram on the floor."

"The floor? They expect me to show up lying on my *back*?" His voice was thick with rage.

I shivered. My bright idea. A pentagram was a prison for demons. Why? I'd thought of the five points of a pentagram, and the five points of a spread-eagled man . . .

"Well?"

"I know, it doesn't make sense. Would you go away now, please?"

He stared. "You *have* forgotten a lot." Slowly and patiently, as to a child, he began to explain the implications of calling up a demon.

I listened. Fear and sick hopelessness rose in me until the concrete walls seemed to blur. "I am in peril of my immortal soul—" This was something I'd never considered, except academically. Now it was worse than that. To hear the demon talk, my soul was already lost. It had been lost since the moment I used the correct spell. I tried to hide my fear, but that was hopeless. With those enormous nostrils he must have smelled it.

He finished, and grinned as if inviting comment.

I said, "Let's go over that again. I only get one wish."

"Right."

"If you don't like the wish I've got to choose another."

"Right."

"That doesn't seem fair."

"Who said anything about fair?"

"—Or traditional. Why hasn't anyone heard about this deal before?"

"This is the standard deal, Jack. We used to give a better deal to some of the marks. The others didn't have time to talk because of that twenty-four-hour clause. If they wrote anything down we'd alter it. We have power over written things which mention us."

"That twenty-four-hour clause. If I haven't taken my wish in twenty-four hours, you'll leave the pentagram and take my soul anyway?"

"That's right."

"And if I do use the wish, you have to remain in the pentagram until my wish is granted, or until twenty-four hours are up. Then you teleport to Hell to report same, and come back for me immediately, reappearing in the pentagram."

"I guess teleport's a good word. I vanish and reappear. Are you getting bright ideas?"

"Like what?"

"I'll make it easy on you. If you erase the pentagram I can appear anywhere. You can erase it and draw it again somewhere else, and I've got to appear inside it."

A question hovered on my tongue. I swallowed it and asked another. "Suppose I wished for immortality?"

"You'd be immortal for what's left of your twenty-four hours." He grinned. His teeth were coal black. "Better hurry. Time's running out."

Time, I thought. Okay. All or nothing.

"Here's my wish. Stop time from passing outside of me."

"Easy enough. Look at your watch."

I didn't want to take my eyes off him, but he just exposed his black teeth again. So—I looked down.

There was a red mark opposite the minute hand on my Rolex. And a black mark opposite the hour hand.

The demon was still there when I looked up, still spread-eagled against the wall, still wearing that knowing grin. I moved around him, waved my hand before his face. When I touched him he felt like marble.

Time had stopped, but the demon had remained. I felt sick with relief.

The second hand on my watch was still moving. I had expected nothing less. Time had stopped for me—for twenty-four hours of interior time. If it had been exterior time I'd have been safe—but of course that was too easy.

I'd thought my way into this mess. I should be able to think my way out, shouldn't I?

I erased the pentagram from the wall, scrubbing until every trace was gone. Then I drew a new one, using a flexible metal tape to get the lines as straight as possible, making it as large as I could get it in the confined space. It was still only two feet across.

I left the basement.

I knew where the nearby churches were, though I hadn't been to one in too long. My car wouldn't start. Neither would my roommate's motorcycle. The spell which enclosed me wasn't big enough. I walked to a Mormon temple three blocks away.

The night was cool and balmy and lovely. City lights blanked out the stars, but there was a fine werewolf's moon hanging way above the empty lot where the Mormon temple should have been.

I walked another eight blocks to find the B'nai B'rith synagogue and the All Saints church. All I got out of it was exercise. I found empty lots. For me, places of worship didn't exist.

I prayed. I didn't believe it would work, but I prayed. If I wasn't heard was it because I didn't expect to be? But I was beginning to feel that the demon had thought of everything, long ago.

What I did with the rest of that long night isn't important. Even to me it didn't feel important. Twenty-four hours, against eternity? I wrote a fast outline on my experiment in demon raising, then tore it up. The demons would only change it. Which meant that my thesis was shot to hell, whatever happened. I carried a real but rigid Scotch terrier into Professor Pauling's room and posed it on his desk. The old tyrant would get a surprise when he looked up. But I spent most of the night outside, walking, looking my last on the world. Once I reached into a police car and flipped the siren on, thought about it, and flipped it off again. Twice I dropped into restaurants and ate someone's order, leaving money which I wouldn't need, paper-clipped to notes which read "The Shadow Strikes."

The hour hand had circled my watch twice. I got back to the basement at twelve-ten, with the long hand five minutes from brenschluss.

That hand seemed painted to the face as I waited. My candles had left a peculiar odor in the basement, an odor overlaid with the stink of demon and the stink of fear. The demon hovered against the wall, no longer in a pentagram, trapped halfway through a wide-armed leap of triumph.

I had an awful thought.

Why had I believed the demon? Everything he'd said might have been a lie. And probably was! I'd been tricked into accepting a gift from the devil! I stood up, thinking furiously—I'd already accepted the gift, but—

The demon glanced to the side and grinned wider when he saw the chalk lines gone. He nodded at me, said, "Back in a flash," and was gone.

I waited. I'd thought my way into this, but—

A cheery bass voice spoke out of the air. "I knew you'd move the pentagram. Made it too small for me, didn't you? Tsk, tsk. Couldn't you guess I'd change my size?"

There were rustlings, and a shimmering in the air. "I know it's here somewhere. I can feel it. Ah."

He was back, spread-eagled before me, two feet tall and three feet off the ground. His black know-it-all grin disappeared when he saw the pentagram wasn't there. Then—he was seven inches tall, eyes bugged in surprise, yelling in a contralto voice. "Whereinhell's the—"

He was two inches of bright red toy soldier. "—Pentagram?" he squealed.

I'd won. Tomorrow I'd get to a church. If necessary, have somebody lead me in blindfold.

He was a small red star.

A buzzing red housefly.

Gone.

It's odd, how quickly you can get religion. Let one demon tell you you're damned . . . Could I really get into a church? Somehow I was sure I'd make it. I'd gotten this far; I'd outthought a demon.

Eventually he'd look down and see the pentagram. Part of it was in plain sight. But it wouldn't help him. Spread-eagled like that, he couldn't reach it to wipe it away. He was trapped for eternity, shrinking toward the infinitesimal but doomed never to reach it, forever trying to appear inside a pentagram which was forever too small. I had drawn it on his bulging belly.

MARTIN GARDNER

No-Sided Professor

Dolores—a tall, black-haired striptease at Chicago's Purple Hat Club—stood in the center of the dance floor and began the slow gyrations of her Cleopatra number, accompanied by soft Egyptian music from the Purple Hatters. The room was dark except for a shaft of emerald light that played over her filmy Egyptian costume and smooth, voluptuous limbs.

A veil draped about her head and shoulders was the first to be removed. Dolores was in the act of letting it drift gracefully to the floor when suddenly a sound like the firing of a shotgun came from somewhere above and the nude body of a large man dropped head first from the ceiling. He caught the veil in midair with his chin and pinned it to the floor with a dull thump.

Pandemonium reigned.

Jake Bowers, the master of ceremonies, yelled for lights and tried to keep back the crowd. The club's manager, who had been standing by the orchestra watching the floor show, threw a tablecloth over the crumpled figure and rolled it over on its back.

The man was breathing heavily, apparently knocked unconscious by the blow on his chin, but otherwise unharmed. He was well over fifty, with a short, neatly trimmed red beard and moustache and a completely bald head. He was built like a professional wrestler.

With considerable difficulty three waiters succeeded in transporting him to the manager's private office in the back, leaving a roomful of bewildered, near-hysterical men and women gaping at the ceiling and each other, and arguing heatedly about the angle and manner of the man's fall. The only hypothesis with even a slight suggestion of sanity was that he had been tossed high into the air from somewhere on the side of the dance floor. But no one saw the tossing. The police were called.

Meanwhile, in the back office the bearded man recovered consciousness. He insisted that he was Dr. Stanislaw Slapenarski, professor of mathematics at the University of Warsaw, and at present a visiting lecturer at the University of Chicago.

Before continuing this curious narrative, I must pause to confess that I was not an eyewitness of the episode just described, having based my account on interviews with the master of ceremonies and several waiters. However, I did participate in a chain of remarkable events which culminated in the professor's unprecedented appearance.

These events began several hours earlier when members of the Moebius Society gathered for their annual banquet in one of the private dining rooms on the second floor of the Purple Hat Club. The Moebius Society is a small, obscure Chicago organization of mathematicians working in the field of topology, one of the youngest and most mysterious of the newer branches of transformation mathematics. To make clear what happened during the evening, it will be necessary at this point to give a brief description of the subject matter of topology.

Topology is difficult to define in nontechnical terms. One way to put it is to say that topology studies the mathematical properties of an object which remain constant regardless of how the object is distorted.

Picture in your mind a doughnut made of soft pliable rubber that can be twisted and stretched as far as you like in any direction. No matter how much this rubber doughnut is distorted (or "transformed" as mathematicians prefer to say), certain properties of the doughnut will remain unchanged. For example, it will always retain a hole. In topology the doughnut shape is called a "torus." A soda straw is merely an elongated torus, so—from a topological point of view—a doughnut and a soda straw are identical figures.

Topology is completely disinterested in quantitative measurements. It is concerned only with basic properties of shape which are unchanged throughout the most radical distortions possible without breaking off pieces of the object and sticking them on again at other spots. If this breaking off were permitted, an object of a given structure could be transformed into an object of any other type of structure, and all original properties would be lost. If the reader will reflect a moment he will soon realize that topology studies the most primitive and fundamental mathematical properties that an object can possess.[1]

A sample problem in topology may be helpful. Imagine a torus (doughnut) surface made of thin rubber like an inner tube. Now imagine a small hole in the side of this torus. Is it possible to turn the torus inside out through this hole, as you might turn a balloon inside out? This is not an easy problem to solve in the imagination.

Although many mathematicians of the eighteenth century wrestled with isolated topological problems, one of the first systematic works in the field was done by August Ferdinand Moebius, a German astronomer who taught at the University of Leipzig during the first half of the last century. Until the time of Moebius it was believed that any surface, such as a piece of paper, had two sides. It was the German astronomer who made the disconcerting discovery that if you take a strip of paper, give it a single half-twist, then paste the ends together, the result is a "unilateral" surface—a surface with only one side!

If you will trouble to make such a strip (known to topologists as the "Moebius surface") and examine it carefully, you will soon discover that the strip actually does consist of only one continuous side and of one continuous edge.

It is hard to believe at first that such a strip can exist, but there it is—a visible, tangible thing that can be constructed in a moment. And it has the indisputable property of one-sidedness, a property it cannot lose no matter how much it is stretched or how it is distorted.[2]

1 The reader who is interested in obtaining a clearer picture of this new mathematics will find excellent articles on topology in the *Encyclopaedia Britannica* (Fourteenth Edition) under *Analysis Situs;* and under *Analysis Situs* in the *Encyclopedia Americana.* There also are readable chapters on elementary topology in two recent books—*Mathematics and the Imagination* by Kasner and Newman, and *What Is Mathematics?* by Courant and Robbins. Slapenarski's published work has not yet been translated from the Polish.

2 The Moebius strip has many terrifying properties. For example, if you cut the strip in half lengthwise, cutting down the center all the way around, the result is not two strips, as might be expected, but one single large strip. But if you begin cutting

But back to the story. As an instructor in mathematics at the University of Chicago with a doctor's thesis in topology to my credit, I had little difficulty in securing admittance into the Moebius Society. Our membership was small—only twenty-six men, most of them Chicago topologists but a few from universities in neighboring towns.

We held regular monthly meetings, rather academic in character, and once a year on November 17 (the anniversary of Moebius's birth) we arranged a banquet at which an outstanding topologist was brought to the city to act as a guest speaker.

The banquet always had its less serious aspects, usually in the form of special entertainment. But this year our funds were low and we decided to hold the celebration at the Purple Hat, where the cost of the dinner would not be too great and where we could enjoy the floor show after the lecture. We were fortunate in having been able to obtain as our guest the distinguished Professor Slapenarski, universally acknowledged as the world's leading topologist and one of the greatest mathematical minds of the century.

Dr. Slapenarski had been in the city several weeks giving a series of lectures at the University of Chicago on the topological aspects of Einstein's theory of space. As a result of my contacts with him at the university, we became good friends and I had been asked to introduce him at the dinner.

We rode to the Purple Hat together in a taxi, and on the way I begged him to give me some inkling of the content of his address. But he only smiled inscrutably and told me, in his thick Polish accent, to wait and see. He had announced his topic as "The No-Sided Surface"—a topic which had aroused such speculation among our members that Dr. Robert Simpson of the University of Wisconsin wrote he was coming to the dinner, the first meeting that he had attended in over a year.[3]

Dr. Simpson is the outstanding authority on topology in the

a third of the way from the side, cutting twice around the strip, the result is one large and one small strip, interlocked. The smaller strip can then be cut in half to yield a single large strip, still interlocked with the other large strip. These weird properties are the basis of an old magic trick with cloth, known to the conjuring profession as the "Afghan bands."

3 Dr. Simpson later confided to me that he had attended the dinner not to hear Slapenarski but to see Dolores.

Middle West and the author of several important papers on topology and nuclear physics in which he vigorously attacks several of Slapenarski's major axioms.

The Polish professor and I arrived a little late. After introducing him to Simpson, then to our other members, we took our seats at the table and I called Slapenarski's attention to our tradition of brightening the banquet with little topological touches. For instance, our napkin rings were silver-plated Moebius strips. Doughnuts were provided with the coffee, and the coffee itself was contained in specially designed cups made in the shape of "Klein's bottle."[4]

After the meal we were served Ballantine's ale, because of the curious trademark,[5] and pretzels in the shapes of the two basic "trefoil" knots.[6] Slapenarski was much amused by these details and even made several suggestions for additional topological curiosities, but the suggestions are too complex to explain here.

After my brief introduction, the Polish doctor stood up, acknowledged the applause with a smile, and cleared his throat. The room instantly became silent. The reader is already familiar with the professor's appearance—his portly frame, reddish beard, and polished pate—but it should be added that there was something in the expression of his face that suggested that he had matters of considerable import to disclose to us.

It would be impossible to give with any fullness the substance of Slapenarski's brilliant, highly technical address. But the gist of it

4 Named after Felix Klein, a brilliant German mathematician, Klein's bottle is a completely closed surface, like the surface of a globe, but without inside or outside. It is unilateral like a Moebius strip, but unlike the strip it has no edges. It can be bisected in such a way that each half becomes a Moebius surface. It will hold a liquid. Nothing frightful happens to the liquid.

5 This trademark is a topological manifold of great interest. Although the three rings are interlocked, no *two* rings are interlocked. In other words, if any one of the rings is removed, the other two rings are completely free of each other. Yet the three together cannot be separated.

6 The trefoil knot is the simplest form of knot that can be tied in a closed curve. It exists in two forms, one a mirror image of the other. Although the two forms are topologically identical, it is impossible to transform one into the other by distortion, an upsetting fact that has caused topologists considerable embarrassment. The study of the properties of knots forms an important branch of topology, though very little is understood as yet about even the simplest knots.

was this. Ten years ago, he said, he had been impressed by a statement of Moebius, in one of his lesser known treatises, that there was no theoretical reason why a surface could not lose *both* its sides—to become in other words, a "nonlateral" surface.

Of course, the professor explained, such a surface was impossible to imagine, but so is the square root of minus one or the hypercube of fourth-dimensional geometry. That a concept is inconceivable has long ago been recognized as no basis for denying either its validity or usefulness in mathematics and modern physics.

We must remember, he added, that even the one-sided surface is inconceivable to anyone who has not seen and handled a Moebius strip. And many persons with well-developed mathematical imaginations are unable to understand how such a strip can exist even when they have one in hand.

I glanced at Dr. Simpson and thought I detected a skeptical smile curving the corners of his mouth.

Slapenarski continued. For many years, he said, he had been engaged in a tireless quest for a no-sided surface. On the basis of analogy with known types of surfaces he had been able to analyze many of the properties of the no-sided surface. Finally one day—and he paused here for dramatic emphasis, sweeping his bright little eyes across the motionless faces of his listeners—he had actually succeeded in constructing a no-sided surface.

His words were like an electric impulse that transmitted itself around the table. Everyone gave a sudden start and shifted his position and looked at his neighbor with raised eyebrows. I noticed that Simpson was shaking his head vigorously. When the speaker walked to the end of the room where a blackboard had been placed, Simpson bent his head and whispered to the man on his left, "It's sheer nonsense. Either Slappy has gone completely mad or he's playing a deliberate prank on all of us."

I think it had occurred to the others also that the lecture was a hoax because I noticed several were smiling to themselves while the professor chalked some elaborate diagrams on the blackboard.

After a somewhat involved discussion of the diagrams (which I was wholly unable to follow) the professor announced that he would conclude his lecture by constructing one of the simpler forms of the no-sided surface. By now we were all grinning at each other. Dr. Simpson's face had more of a smirk than a grin.

Slapenarski produced from his coat pocket a sheet of pale blue paper, a small pair of scissors, and a tube of paste. He cut the paper into a figure that had a striking resemblance, I thought, to a paper doll. There were five projecting strips or appendages that resembled a head and four limbs. Then he folded and pasted the sheet carefully. It was an intricate procedure. Strips went over and under each other in an odd fashion until finally only two ends projected. Dr. Slapenarski then applied a dab of paste to one of these ends.

"Gentlemen," he said, holding up the twisted blue construction and turning it about for all to see, "you are about to witness the first public demonstration of the Slapenarski surface."

So saying, he pressed one of the projecting ends against the other.

There was a loud pop, like the bursting of a light bulb, and the paper figure vanished in his hands!

For a moment we were too stunned to move, then with one accord we broke into laughter and applause.

We were convinced, of course, that we were the victims of an elaborate joke. But it had been beautifully executed. I assumed, as did the others, that we had witnessed an ingenious chemical trick with paper—paper treated so it could be ignited by friction or some similar method and caused to explode without leaving an ash.

But I noticed that the professor seemed disconcerted by the laughter, and his face was beginning to turn the color of his beard. He smiled in an embarrassed way and sat down. The applause subsided slowly.

Falling in with the preposterous mood of the evening we all clustered around him and congratulated him warmly on his remarkable discovery. Then the man in charge of arrangements reminded us that a table had been reserved below so those interested in remaining could enjoy some drinks and see the floor show.

The room gradually cleared of everyone except Slapenarski, Simpson, and myself. The two famous topologists were standing in front of the blackboard. Simpson was smiling broadly and gesturing toward one of the diagrams.

"The fallacy in your proof was beautifully concealed, doctor," he said. "I wonder if any of the others caught it."

The Polish mathematician was not amused.

"There is no fallacy in my proof," he said impatiently.

"Oh, come now, doctor," Simpson said. "Of course there's a fallacy." Still smiling, he touched a corner of the diagram with his thumb. "These lines can't possibly intersect within the manifold. The intersection is somewhere out here." He waved his hand off to the right.

Slapenarski's face was growing red again.

"I tell you there is no fallacy," he repeated, his voice rising. Then slowly, speaking his words carefully and explosively, he went over the proof once more, rapping the blackboard at intervals with his knuckles.

Simpson listened gravely, and at one point interrupted with an objection. The objection was answered. A moment later he raised a second objection. The second objection was answered. I stood aside without saying anything. The discussion was too far above my head.

Then they began to raise their voices. I have already spoken of Simpson's long-standing controversy with Slapenarski over several basic topological axioms. Some of these axioms were now being brought into the argument.

"But I tell you the transformation is not bicontinuous and therefore the two sets cannot be homeomorphic," Simpson shouted.

The veins on the Polish mathematician's temples were standing out in sharp relief. "Then suppose you explain to me why my manifold vanished," he yelled back.

"It was nothing but a cheap conjuring trick," snorted Simpson. "I don't know how it worked and I don't care, but it certainly wasn't because the manifold became nonlateral."

"Oh it wasn't, wasn't it?" Slapenarski said between his teeth. Before I had a chance to intervene he had sent his huge fist crashing into the jaw of Dr. Simpson. The Wisconsin professsor groaned and dropped to the floor. Slapenarski turned and glared at me wildly.

"Get back, young man," he said. As he outweighed me by at least one hundred pounds, I got back.

Then I watched in horror what was taking place. With insane fury still flaming on his face, Slapenarski had knelt beside the limp body and was twisting the arms and legs into fantastic knots. He was, in fact, folding the Wisconsin topologist as he had folded his piece of paper! Suddenly there was a small explosion, like the backfire of a car, and under the Polish mathematician's hands lay the collapsed clothing of Dr. Simpson.

Simpson had become a nonlateral surface.

Slapenarski stood up, breathing with difficulty and holding in his hands a tweed coat with vest, shirt, and underwear top inside. He opened his hands and let the garments fall on top of the clothing on the floor. Great drops of perspiration rolled down his face. He muttered in Polish, then beat his fists against his forehead.

I recovered enough presence of mind to move to the entrance of the room, and lock the door. When I spoke my voice sounded weak. "Can he . . . be brought back?"

"I do not know, I do not know," Slapenarski wailed. "I have only begun the study of these surfaces—only just begun. I have no way of knowing where he is. Undoubtedly it is one of the higher dimensions, probably one of the odd-numbered ones. God knows which one."

Then he grabbed me suddenly by my coat lapels and shook me so violently that a bridge on my upper teeth came loose. "I must go to him," he said. "It is the least I can do—the very least."

He sat down on the floor and began interweaving arms and legs.

"Do not stand there like an idiot!" he yelled. "Here—some assistance."

I adjusted my bridge, then helped him twist his right arm under his left leg and back around his head until he was able to grip his right ear. Then his left arm had to be twisted in a somewhat similar fashion. "Over, not under," he shouted. It was with difficulty that I was able to force his left hand close enough to his face so he could grasp his nose.

There was another explosive noise, much louder than the sound made by Simpson, and a sudden blast of cold wind across my face. When I opened my eyes I saw the second heap of crumpled clothing on the floor.

While I was staring stupidly at the two piles of clothing there was a muffled sort of "pfft" sound behind me. I turned and saw Simpson standing near the wall, naked and shivering. His face was white. Then his knees buckled and he sank to the floor. There were vivid red marks at various places where his limbs had been pressed tightly against each other.

I stumbled to the door, unlocked it, and started down the stairway after a strong drink—for myself. I became conscious of a violent hubbub on the dance floor. Slapenarski had, a few moments earlier, completed his sensational dive.

In a back room below I found the other members of the Moebius Society and various officials of the Purple Hat Club in noisy, incoherent debate. Slapenarski was sitting in a chair with a tablecloth wrapped around him and holding a handkerchief filled with ice cubes against the side of his jaw.

"Simpson is back," I said. "He fainted but I think he's okay."

"Thank heavens," Slapenarski mumbled.

The officials and patrons of the Purple Hat never understood, of course, what happened that wild night, and our attempts to explain made matters worse. The police arrived, adding to the confusion.

We finally got the two professors dressed and on their feet, and made an escape by promising to return the following day with our lawyers. The manager seemed to think the club had been the victim of an outlandish plot, and threatened to sue for damages against what he called the club's "refined reputation." As it turned out, the incident proved to be magnificent word-of-mouth advertising and eventually the club dropped the case. The papers heard the story, of course, but promptly dismissed it as an uncouth publicity stunt cooked up by Phanstiehl, the Purple Hat's press agent.

Simpson was unhurt, but Slapenarski's jaw had been broken. I took him to Billings Hospital, near the university, and in his hospital room late that night he told me what he thought had happened. Apparently Simpson had entered a higher dimension (very likely the fifth) on level ground.

When he recovered consciousness he unhooked himself and immediately reappeared as a normal three-dimensional torus with outside and inside surfaces. But Slapenarski had worse luck. He had landed on some sort of slope. There was nothing to see—only a gray, undifferentiated fog on all sides—but he had the distinct sensation of rolling down a hill.

He tried to keep a grip on his nose but was unable to maintain it. His right hand slipped free before he reached the bottom of the incline. As a result, he unfolded himself and tumbled back into three-dimensional space and the middle of Dolores's Egyptian routine.

At any rate that was the way Slapenarski had it figured out.

He was several weeks in the hospital, refusing to see anyone until the day of his release, when I accompanied him to the Union Station. He caught a train to New York and I never saw him again. He died a few months later of a heart attack in Warsaw. At present Dr. Simpson is in correspondence with his widow in an attempt to obtain his notes on nonlateral surfaces.

Whether these notes will or will not be intelligible to American topologists (assuming we can obtain them) remains to be seen. We have made numerous experiments with folded paper, but so far have produced only commonplace bilateral and unilateral surfaces. Although it was I who helped Slapenarski fold himself, the excitement of the moment apparently erased the details from my mind.

But I shall never forget one remark the great topologist made to me the night of his accident, just before I left him at the hospital.

"It was fortunate," he said, "that both Simpson and I released our right hand before the left."

"Why?" I asked.

Slapernarski shuddered.

"We would have been inside out," he said.

WILLIAM F. ORR

Euclid Alone

1

The elevator was passing the tenth floor when Dr. Donald Lucus started from his reverie into the panic of embarrassment he always felt when he had passed his floor unknowingly. His mind shifted with reluctance from one program to another as it was drawn to the demand for a decision by the illusion of a sudden decrease in gravity that indicated the elevator would stop on the eleventh floor. He would have approximately ten seconds to decide. He could stay on this elevator, possibly all the way to the twenty-fifth floor, and get off at six on the way down, wasting a large but certain amount of his time, resulting in only a brief moment of embarrassment before unknown engineers on the top floor, who would realize his foolish mistake when he remained in the elevator, who would smile and wonder why this absentminded old fellow hadn't been retired by the Institute yet. Or he could get off now, at eleven, and wait for the next down elevator to take him to his own floor. Again, it would be apparent to the secretary at Genetics that he had gone by his floor, and he would be aware of her, sitting behind him at her desk, thinking he had been a section head much too long and ought to be replaced. As an alternative, he could resort to subterfuge and walk down the hall to knock on the door of an empty office, pretending that he had legitimate business there.

But he had neither the will nor the energy to act out such a pantomime, when his mind needed urgently to be occupied with another problem, which it had been drawn away from by this trivial face-saving decision making. In the end, he decided that getting off at eleven would waste less time than any other course of action, and that was, after all, the most important consideration.

As it was, he made this decision too late to avoid another embarrassment, that of being hammered on both sides by the elevator doors as he stepped out, while the secretary of Genetics looked up and smiled. An automatic mumbled "Pardon me" crossed his lips as he pulled free of the gentle visegrip and turned self-consciously to push the down button.

Only then could he relax and let his mind sink back to the complex but ordered patterns of proof which were its true medium. And that order had form quite as real as the world of elevator buttons and social ineptitude, of old men and young men and publishing deadlines and hiring policies. There was an intricate structure made entirely of straight lines in a plane that intersected in named points and formed identifiable triangles, all related to one another by a carefully chosen pattern of congruences, similarities, and equal sides and angles.

This structure had been built very carefully to its present state by a process of repetitive partial construction. All the way along the freeway he had occupied himself with the task of mentally rebuilding the structure which he had spent almost the whole night examining. He would begin each time at the same starting point, building one line at a time, noting each label, each equality and similarity, until the structure reached a degree of complexity, as it did each time, that exceeded his ability to assimilate new information and which became manifest in the sudden complete loss of a necessary fact. At this point the only possibility was to begin again at the foundation. Each time, he got a little farther in the proof, and while it might seem at first a most inefficient method of construction, it would eventually result, not only in a completed proof, but also, and more importantly, in a complete intuitive familiarity with that proof, both in its overall conception and in all its particulars.

It was a learning method that Dr. Lucus could remember having employed successfully for over forty years, and even his loss of mental agility resulted only in a longer amount of time spent with any one proof and not in any loss of total comprehension once the process was completed.

He had taken temporary comfortable refuge in the beauty and symmetry of this proof, which he could do now, shutting out all thought of the threat, the horror of its ultimate implications. The four cups of coffee kept him awake and uncomfortably numb to the eventual attack on his one sure foothold on reality. He stepped into the elevator and forcibly turned his thoughts from what he must do in the next few hours. For he was not certain what he must, could, or wanted to do. Too many roads were open to him, and they all seemed to lead to eventual dead ends. But that was only, as Hans used to tell him, his own lack of imagination, his lack of initiative to build his own road. So be it. As he began once more to piece together his elusive triangles, he was aware more than anything of the face of the secretary of Genetics, frozen in the last narrow inch between the closing elevator doors, smiling at an old, absent-minded man who held in his mind the cursed flame of destruction of this whole temple of reason.

If Ruth was aware of the worn, drugged look of his eyes and face when he entered the Math office, she did not betray this. She turned from her typing, pulled her glasses down on her nose, and regarded him with what she supposed was a friendly smile, as she always did. After the obligatory good-mornings, he paused, trying to force efficiency into his reflexes, confused over what were to be his instructions to her.

"Uh, Ruth," he began at length, "Ruth, would you get the Director on the line for me?"

"The Director, sir?"

She was not, in fact, asking for confirmation, only registering surprise. Dr. Lucus seldom had any contact with the Director, except at executive board meetings. When he did contact the Director, it was always through interoffice memo, and Ruth had shown her surprise at this breach of tradition before she had a chance to check herself. One telephoned subordinates and colleagues. One wrote to superiors.

"Uh, yes, Ruth, it's rather important."

Was there something else to say to Ruth? He had to call Publications, but that could wait until after he got the go-ahead from the Director.

"Oh, uh, Ruth, is the mail in yet?"

"No, sir, but there was a telegram. It's on your desk." She was impatient to get back to her typing, impatient with his slow talk and his hesitation. But she would not return to work until he left the room. Office etiquette was her one comfort in this job.

"Oh, uh, thank you, Ruth." He turned to his office. At the door he paused a moment, as she said, in accordance with custom, "Would you like me to bring you a cup of coffee?"

He had anticipated the question, so his answer was immediate—in fact, clumsily abrupt. "No, thank you, Ruth."

As the door swung shut, the steady patter of her machine resumed. She would type one page to allow him time to take off his coat and get settled. She would sip her own lukewarm cup of coffee, which she nursed for three hours every morning, and then turn to the telephone. Her wrinkled face would show no sign of the joy she felt at hearing her own crisp, stiff voice conducting business efficiently and properly.

His hand was strangely empty as he took off his coat. He had left his briefcase in the car.

"Damn," he muttered. That meant he would have to go back for it during his lunch hour. He would definitely need to have the papers in it before he saw the Director. In fact, he had wanted to review Professor David's paper this morning and to check the mental construction he had prepared in the car. In any case, he would need the paper itself before he could run a computer check on the validity of the proof.

All these trivial irritations made it even more difficult to see through the haze of a sleepless night that he was caught up in something historic, in something frightening. He was not made for scrapping a lifetime of firmly held beliefs in a day, as Hans was. He was not made to be forced suddenly and rudely into a crucial position of responsibility. Hans could handle that sort of thing; he could not. Hans could submerge himself in madness and come up smiling, happy and sane. But for Donald Lucus madness, if it came, would be the end. He rested a hand on Hans's sculpture as he sat at the desk. In the outer office, the sound of typing stopped.

There was a thin wire human figure suspended inside a cage, which was formed by the edges of an irregular icosahedron, slightly skewed on its axis. Two edges of the polyhedron were broken, and the figure was falling, one hand stretched out vainly toward an edge, a bar of the cage. Its mouth was open.

"This," the artist had said to him years before over a game of Go, "this is you, Don. Not now. This is you at sixty-five. This is you and your twenty-faced monster and your quintic equation. I want you to save it for your old age and then tell me if I'm right."

He had clicked the cigarette holder between his teeth and grinned that diabolical smile that always dared you to guess whether he was joking or serious.

"Look at it, Don. And when you finally recognize yourself, write and let me know."

The telegram was brief and clear.

> DON.
> HANS DIED OF A STROKE FRIDAY. BETH ASKED ME TO LET YOU KNOW. FUNERAL WEDNESDAY AT ONE ST. PAULS, CINCINNATI. WILL NOT BE THERE AS THINK THAT WOULD BE BETTER FOR BETH. HOPE YOU CAN THOUGH. MUCH LOVE.
>
> MARY

"Dr. Lucus, I have the Director's secretary on the line. The Director will not be in until ten this morning. His schedule is full today, but I can make an appointment for you Tuesday afternoon."

Tuesday afternoon. He set the telegram down and covered his eyes to think. Tuesday afternoon. He could make an appointment for Tuesday afternoon, and that would give him another day to relax, another day before he really had to do anything about the situation. But no, that was impossible. He couldn't put it off, and he couldn't *be* put off. Of course, he must see the Director immediately.

"Dr. Lucus?"

"Uh, yes, yes, Ruth. Uh, thank you, but I really must talk to the Director as soon as he comes in. It's . . . it's quite important. Would you leave a message for him to call me? It's very high priority, Ruth." He thought high priority sounded better than urgent, more professional. It was a term the Director would probably use.

"Yes, sir. Will that be all?" There was only a faint sign of reproach in her voice for this break with tradition.

"No, that's all, Ruth. Thank you."

But there was more. There was much more he had to do before lunch.

"Wait! Ruth? Ruth, would you get Publications on the phone? I want to speak to Jack Hudson. That's right. Thank you, Ruth."

He sat frozen behind his desk. He had to talk to Hudson and stop today's mailing, to recall any copies that had already been sent out, to hold them until someone could make a final decision. Someone. He should run a check on the computer, and a projection too. Ordering these things in his mind was a difficult task. There were too many factors to tell what to do first. The whole pattern of his schedule was torn, and he had left his damn briefcase with that damn paper in the car. Construct angle $F'\ G'\ H$ = angle $GG'B$. Then, if AJ is dropped perpendicular to BG from A, $BJ = AJ$ and $BG = F'H$. Thus triangle ABG is congruent to . . . is congruent to . . . He rose, rushed to the blackboard, and began drawing furiously, attacking that hideous proof directly, headlong. He *must* find a fallacy. It *must* be false. He drew in three colors of chalk, erasing and redrawing segments in new proportions, stepping back across the room to view his diagram from a distance, making quick notations on the back of the piece of yellow paper on his desk, pacing the room jerkily and returning to the board to scowl, erase, and redraw. He hardly noticed it when Ruth buzzed to tell him Hudson was on the line. He strode to the desk, one eye on the board, and surprised himself by his handling of the situation.

Was the autumn number of the *Quarterly Mathematics Publications of the Federal Basic Research Institute* ready for mailing? Good, hold it until further notice. No, no serious problem. A rather important error that would have to be corrected. A paper might have to be removed. Had any copies been sent out? Five review copies had gone out earlier. Please have them recalled. As soon as possible, yes. Lucus himself would write to the *American Mathematical Monthly*. No, it wasn't a serious problem. It would be rather difficult to explain to a nonmathematician. Thank you very much for your cooperation, Mr. Hudson. Terribly sorry to cause your department all this bother. Yes, thank you. Good-bye, Mr. Hudson.

It took only a moment at the blackboard to regain his balance, to recover his position. His construction started at the bottom, spread out on both sides, and then began climbing upward, just as it had in David's paper. Like some sort of tower. That's what Hans would call it, if he were here. A tower of matchsticks, something like that. Then he would insist that he was going to do a painting of it. And he might, in fact. It was rather attractive, quite nicely symmetric. The whole picture had a neat look of innocence about it, as if it were nothing more than a new proof of the Pythagorean theorem, for example. Hans would call it "The Tower That Demolished the Tower" or something like that and find the irony of it hilariously funny. The destruction of centuries of mathematical thought would mean nothing to him. It was a joke. A joke on Lucus, a confirmation of everything Hans had said over those interminable Tuesday-night games of Go.

As the tower neared its peak, it became increasingly obvious to Dr. Lucus that the proof was correct, that there was no fallacy. After six times through it, he could no longer tell himself he was not following it well enough, that he would see the obvious hole in the logic the next time through. David had been meticulous, he had left out no steps. His paper was densely written and quite thorough. Euclidean geometry was not Donald Lucus's field. He was not used to its methods of proof. But by now he could feel each lemma and corollary of David's theorem in his guts. He knew it was true. Only a computer confirmation remained.

He called the computer office and arranged for some time that evening, asked for three tapes to be sent to his office: first, CONPROOF2: Confirmation of the consistency of a proof in a mathematical axiom system given as a subroutine; second, EUBERT: Hilbert's axioms for Euclidean geometry, subroutine for CONPROOF2; and finally, LOBACHEVMANN: Lobachevskian and Riemannian geometries, subroutine for use with CONPROOF2.

Then he called the head of the computer division and explained that he wanted to run a social projection later in the week. The man was incredulous—and amused.

"In Math?" he asked.

"Yes," replied Lucus. "And I want the problem and the output to be considered Limited Interest."

The man paused only a second at the other end of the line, his mouth hanging open an inch from the receiver.

"I . . . I'll have to have an okay from the Director on that first, but all right, I'll see what I can keep open for you Friday night."

Lucus thanked him and hung up. His palms were covered with sweat. The use of the term "Limited Interest" had frightened him and impressed on him the seriousness of the thing he was doing. There was no classified research at the FBRI; its fundamental philosophy was one of "basic research in a free and open environment." All the work done in the building or under FBRI grants was published and widely disseminated. However, it occasionally became clear to the Institute officials that certain results could prove dangerous in one way or another if prematurely released to the public or to the scientific community at large. Therefore the code "Limited Interest" had been developed to refer to such work: unclassified, but kept strictly under wraps.

It was nine-thirty, and he had done all he could until he talked to the Director. The tower stood flat against the blackboard, a dead, crystalline, cutting blade of red and blue and orange. Outside the window, a squirrel darted along a branch and vanished down one of its countless customary routes in the maze of almost leafless branches. The Institute was built into the side of a hill, so that Dr. Lucus's office, which was on the sixth floor if seen from the front, actually appeared to be no higher than the third floor. He had a peaceful view of grass and sky, held fast by the swift, layered lattice of branches. Often he felt that he did his best thinking while he was standing here, running his eyes peacefully along the branching lines, like one of Kaufmann's illustrations in *Graphs, Dynamic Programming, and Finite Games*. But today he could not think in leisure. The soap-white walls that rose to enfold his world were too close now, and he was trapped, trapped and falling.

He did have one thing left to do, although he did not feel like seeing Ruth. It must be done, and it would give his mind something to grasp until he could talk to the Director. He had her come in, and he dictated a polite letter, a bit too long, to the editor of the *AMM,* explaining that an embarrassing error had crept into the fall *Quarterly* and asking that it be returned, so that it could be replaced by a corrected copy.

As Ruth got up to go, her steno pad pushed the yellow paper off his desk to the floor. He bent to pick it up and smoothed it on his desk, staring blankly at the calculations on the back. His watch said 10:05.

At ten-fifteen he still sat, frozen, his eyes open and filled with the erected sword shape plastered flat against the blackboard. It was cutting deep into his retinas, but his mind was suspended in dreamless waking sleep that numbed the wound and held him in inanimate rigid repose. Solidity of metal and wood near him and touching him melted, and only the neat impersonal sword hung above, simple, clear, no longer threatening, neutral now as all things were within the asbestos web which held him.

He had to force his head down, to force his eyes to see the desk, the wrinkled yellow on white metal. To force his arm up, pull back the sleeve, and decide to act. Only then did his focus return, caught by the jittering watch, and then the office took clear familiar shape again around him. And fear returned.

"Ruth? Would you call the Director's office again and see if he's in yet?"

When the Director finally came on the line, he was curt and impatient. He didn't like to be bothered by petty problems; Lucus knew that.

"It's really quite urgent. I don't like to upset your schedule, but I'm afraid it can't wait until tomorrow, sir."

"Well, what is it in Math that you can't handle yourself, Lucus?" His voice was overamplified by the receiver, and there was no comfortable position for it. "If there are problems with funds or payroll, that shouldn't be handled through my office. I should think you would be able to take care of your own distribution of grants."

"Well, no, sir, it's a more important problem than that. It's research that I feel needs . . . uh . . . special attention. I mean there seem to be possible . . . possible dangers in publication of certain discoveries."

"In math, Lucus? You're exaggerating. What sort of research in math could produce . . . uh, dangers? I mean, you're surely getting carried away with your formulas, aren't you?"

"I'd rather not discuss it over the phone, sir. It is of rather . . . of rather Limited Interest."

"Oh?" Lucus could feel the younger man's eyebrows rising. "What sort of 'Limited Interest,' Dr. Lucus?"

"If I could make an appointment, sir . . ."

"My schedule is terribly busy, Dr. Lucus, and I don't see how I can fit in another appointment—unless you will tell me the nature of the problem."

Lucus was not ready for this, not ready to reveal to anyone else the secret that, as far as he knew, he alone shared with Professor Paul David. He had not thought this far. He would have to tell another man the horrible thing that had been discovered, the horrible thing that had lain in wait for discovery all these centuries. His face covered with sweat, his hand sticky against the plastic receiver, he controlled his voice as much as he could and said, "A disproof of Euclid, sir. One of our fundees has produced a proof of the inconsistency of Euclid . . . that Euclid is not true, *cannot* be true . . ."

There was no reply, no sound. He didn't know if the Director was as shocked as he, or if he was incredulous, unable to believe such a thing. Did he perhaps share Hans's sense of the cosmic joke of the whole thing? Was he smiling with the chemist's triumphant smile at the defeat of the abstract theoretician? How would any man react to such knowledge? Lucus decided that the Director did not believe him, that no man could accept such a horrifying conclusion without rigid proof. Surely he himself had spent two days and sleepless nights in the attempt to shake the unshakable conclusion. Finally the voice answered. It was a short answer and made its point perfectly clear.

"Is that all? I think you should be able to clear that up by yourself, Lucus. After all, I don't know much about math, and I don't see why you have to bring it to my attention."

And that was it. He was unimpressed. It meant nothing to him.

"I think, sir, that it is very important that I explain the problem to you in more detail."

"All right, Lucus, all right. Come by at three, will you? I have an important call on the other line. Sorry, I have to hang up. At three."

The sigh of resignation in his voice had been almost theatrical. He hated to be bothered with petty departmental problems.

Dr. Lucus cradled his head in his arms on the desk, shaking uncontrollably with the release of tension, still alone in his fear and in his knowledge.

2

"Don, you're using strategy again. It's the same strategy; it's a textbook play." Hans Kaefig blew a thick puff of smoke at the stilled Go board. "Blurred your vision a little, Donnie. Come on, find a play that doesn't have a proverb to go with it."

Hans leaned back in his chair, hands behind his head, cigarette holder rising out of his bushy gray beard like a radius vector tracing minute burning circles in the air. He closed his eyes tight in a pantomime of cogitation. "What you need to do, Donnie, what you need to do is . . ."

Don smiled and fumbled with his pipe. He knew Hans could go on like this all night, fighting his own eternal battle with rational thought out loud, using Don's career as his battlefield, giving him advice, often self-contradictory, on how to break from the confines of Aristotelian logic and soar like a bird on the soul of his intellect. Or the intellect of his soul. The words varied proportionally to the amount of brandy consumed every Tuesday.

". . . what you need, Donnie, is to state a theorem without a proof—with no hope of a proof. Write a paper, Don, with ten or twenty wild, impossible theorems and lemmas and corollaries—no proofs . . . absurd theorems. I'll help you. I'll give you some ideas, you can rewrite them to sound mathematical. We'll publish them—inside a year someone will have proved half of them, done all the work, but they'll all be called Lucus's theorem or Kaefig's conjecture—and we'll have it made."

"Wouldn't work, Hans. No one would publish them without proof."

"Well, then—we'll publish it as a novel. That's it, a novel. You write the theorems, I'll write the sex. We'll call it *Propositional Calculus,* and start a rumor that it was written under drugs. We can cut the verbs out of all the sentences and make it look all Burroughs-y. That's the way to do mathematics, Don. Get out of the mainstream . . . underground math . . . subversive topology, that should be your field, luv."

"Hans, I appreciate your help with my career—"

"It's only that I pity you—I'm determined to make an artist of you, if you don't make me into a scientist first."

"Now have I tried to do that?"

"Oh, you're subtle, Donnie. You're subtle. And that's what I'm not. You can see my plan of action right away. But you—well, you just leave those books lying around open so I'll sneak a peek. You try to draw me to those dirty pictures: a truncated cube, a stellated dodecawhatsit, two pyramids stuck through each other . . . it's warping my brain. I go to my studio and find my mind all hung up in your simply connected sets and those tragic asymptotic curves—Tantalus damned to approach without reaching forever. What can I do? You have told me a doughnut is a coffee cup, and I have believed you. I used to paint the city and garbage and reality. Now all I know are points in space. I dream each night of being trapped in Königsberg, forever recrossing those bridges, while Euler stands by the river and laughs. Oh, don't deny it, Don. You are slowly turning me into what you are, enveloping me in symbolic logic and set theory. And I keep coming back and asking for more."

"And why do you come back?" asked Don, finishing off his brandy.

"Ah, you force me to say it! You are my muse, Professor Lucus. Without you my art would die. Without you, my dear friend, I would paint only the city and garbage and reality."

"Oh . . . I thought Mary was your muse."

"Mary? Of course not. A muse must be unobtainable, mysterious, the artist's opposite, the soul of what he can never be. You might as well say I am Mary's muse. After all, she did dedicate a quartet to me. Am I flattered? No. We're getting divorced this year, or next year—whenever there's time."

"You're not serious, Hans!" Don hated the terrible uncertainty. In fact, he never did know when Hans was serious about *any* subject.

"Of course I am. It was her idea. Or maybe it was mine. Anyway, we talked it over and thought it would be fun. But now you've sidetracked me. I was explaining your role in my art. Ever since I started playing Go over here, look how I've improved. Look at that sketch I did of your continuous function theorem. The critics love it. They think I'm a genius."

"It's ugly, you know. It's really ugly." As much as he liked Hans, Don had never been able really to appreciate any of his work. He found it childish, simple, and sometimes repulsive. Since Hans had begun basing his things on diagrams in Don's math books, he liked them even less. They seemed to make art lifeless and mathematics unprincipled.

"Yes, it is," agreed Hans. "Very ugly. And you see, before you inspired me I'd never been able to paint anything quite that ugly. I'd tried . . . Lord, I have as good a sense of what is offensive to the eye as any other artist, but I'd never been able to put it down on canvas. I would walk around in the slums and look at the *dreck* in the alleys and think it was ugly. I would eat starch sandwiches at the Automat and wipe my beard with a used napkin and think that was ugly. But then when I looked into your Hocking and Young and saw that wild sphere—I knew I had found it! I knew other men had seen the true vision, and I could learn from them."

"You're crazy, Kaefig," Don intoned, shaking his head.

Hans clicked his teeth against his empty cigarette holder and drew a pack of Camels from the pocket of his Levi jacket. "You've said that ever since college, my dear professor, and it hasn't made it any less true, you smug sane bastard. Let's put away the game and get drunk. I think I've done quite enough to try to save your soul for one night."

And so it went for over seven years, from Don's thirty-eighth birthday deep into his forties, until Hans and Mary finally split up and he moved to the Midwest.

And Don never tired of his friend's harangues, because he knew there was something important there, something he should hear. And so he listened to all the nonsense and rambling, trying to sift out the bit of informational content, the little he could really learn from Hans.

Mary had been hired as conductor of the Denver Symphony, and Don had heard little from either of them since, except for sporadic Christmas cards. He had left teaching to come to the Institute at fifty-five, and there he had remained, sitting—how had Hans put it?—sitting on top of that pile of elephant tusks, lord of what little he surveyed.

A few years ago, a journalist had interviewed Dr. Lucus, because Hans had said he was the only man who could explain his sculpture *Ragtime Band,* that sprawling monstrosity that was the culmination of his fascination with the wild sphere in Hocking and Young and, according to many critics, the culmination of his career.

And now all that remained of those endless games of Go were a couple of Hans's paintings in Donald Lucus's house, the wire sculpture on his desk, and a telegram with a few hasty calculations on the back.

He worked on the program until past his usual lunchtime, carefully cross-referencing the manuals spread out on his desk. The first step had been to write out the entire proof, as well as he could remember it, but with the diagram on the blackboard to help him. This was written in his own private notation, a hybrid of FORTRAN, mathematical symbology, and abbreviated English. The next step was to translate this into symbolic logic, using the special terms and syntax laid out by CONPROOF2 and EUBERT. Not only was it necessary to translate from one code to another; in order to avoid an impossible mass of detail, Lucus also had to augment the Hilbert axioms for Euclidean geometry with statements of all the Euclidean propositions called upon in David's proof. There was a list of these in the supplementary notes on EUBERT, and so he didn't have to worry about coding them, only that he had inserted all the necessary ones and correctly labeled them.

This was only the Euclidean part. For a while, he was afraid it would be necessary to duplicate all this work to program his Lobachevskian and Riemannian checks. But then he discovered a special tie-in in LOBACHEVMANN which would allow him to use the exact same input as was used for EUBERT and have validity checked in both non-Euclidean geometries at the same time. He was familiar enough with the use of the old CONPROOF1, but only in conjunction with such systems as TOPOSPACE and ENSN, which he used constantly in verifying topological proofs. The axiom systems for synthetic geometries had been a complete mystery to him for over thirty years, and what he had relearned in the last three days was hasty and incomplete.

So his office, normally neat to the point of sterility, took on the aspect which it had only a few days a month, those few days of feverish inspiration when he had all the business details of his position out of the way and could allow himself the luxury of creation. Directly in front of him were a programming pad, on which he was writing his final version, and a pile of scratch work. Across the upper part of the desk the three program manuals lay open; to his left was his recent reproduction of David's proof, continuous on the back of last month's budgetary output, and a stack of used scratch paper which contained calculations important enough to be saved; on his right, a well-worn FORTRAN manual on top of the three books which were almost the only customary adornment of the desk: *Webster's New Collegiate Dictionary,* Whittier's *Trilingual Mathematical Encyclopedia,* and a book of Go proverbs. The wire sculpture and desk calendar had been moved to the file cabinet to make room for all the necessary reference material.

By twelve-thirty Ruth had gone to lunch, but the programming was not nearly completed. He had to pull himself away from the pad and pencil almost violently. His hand, his whole body, and a portion of his mind were unwilling or unable to stop writing. Once the trance was broken and he was putting on his coat, it began to frighten him. Surely he worked efficiently in such a hyperactive state, but it was dangerous. He could easily push himself too far, almost unknowingly, uncaringly, if he were allowed to give himself up to the immersing impulse too often. Even as he walked through the outer office, he noticed a stiffness in his legs and neck, an ache in his back and hand, that he had been oblivious to minutes before. Returning to awareness of his body's torture, he found the temporary divorce from objectivity even more frightening, as though it had been imposed not by himself but from the oustide, as though he had been driven too hard by some other being, with little or no concern for his complaints or his safety. He had been abused. And he was tired, very, very tired.

His stomach was feeling upset—from the two cups of coffee he had had with lunch, he supposed—when he returned with his briefcase and David's paper at one-fifteen. He found that once he lowered himself into the swivel chair it was necessary to sit still for several minutes to catch his breath. He knew he needed rest, but there was much more to do before he saw the Director. He tried to weigh the priorities in his mind, to reach a reasonable plan of action, but it was difficult to pin down ideas, and his thoughts were constantly intruded upon by images of congruent triangles and hyperbolic planes. Each attempt to list the tasks of the afternoon and assign time estimates to them was met with frustration, and his ears rang with the faint sound of laughing voices chattering in FORTRAN.

Finally he decided that the only really necessary task was to finish the programming for CONPROOF2 and have it sent to the computer division.

This took about half an hour. The work went much slower than it had in the morning. He constantly found himself looking blankly at his own notes, confusing output statements with axioms, losing his tenuous grasp on the details of David's proof.

When the whole thing was finally sent out to be punched up and compiler-checked on the A50 unit, it was ten minutes to two. He told Ruth to call him at five to three and gratefully laid his head on his folded arms on the desk, not even bothering to darken the room.

At first sleep would not come, only sharp-edged pictures, alternately threatening and soothing: his small house, his books, Mary as she was thirty years ago, when he had thought he might marry. Unrelated images swam about in his mind: a snatch of old rock music, a lemma from the side approximation theorem, the smell of lilacs, and slowly one figure emerged from the mass and began to grow and dominate it all. At first it appeared to be only a smooth, featureless, somewhat metallic topological sphere. But this was only the bottom portion. Above, the figure split into two parts, not so much like a branching tree as like a squid with two plump arms, spread in a gentle flattened circle and coming together again—but not quite. Before the arms met, each of them split into two parts again, a thumb and an index finger, which linked together like a chain—but not quite. Before each finger reached its thumb, it was bifurcated, as was the thumb, and again the two arcs linked to form a chain—but not quite. This process continued infinitely, each step increasing the number of parts geometrically. The result was a figure simple in its construction, frighteningly complex in its final appearance.

"My God in heaven!" exclaimed Hans. "What in merciful hell is that? I never expected to find a book on demonology in your home, Donald."

Don had been at the bar pouring brandy and didn't know what Hans was referring to. By the time he got back to the table, Hans was standing silently, biting the side of his thumb and staring fixedly at the open copy of Hocking and Young's *Topology*.

"That's Alexander's horned sphere," explained Don evenly. For a moment he too felt an uncanny horror at the picture. But it passed, and what was to Hans the image of Satan became only a wild embedding of the 2-sphere in Euclidean 3-space.

"You see," he went on, setting down the drinks, "it's topologically a sphere, but its complement is not the same as the complement of an ordinary sphere. For instance, you could link a circle around it, just like you would around a torus, and it won't come off. In fact, there are an infinite number of ways you can do it."

Hans seemed not to be listening. He did not respond when Don sat down, but continued to bite his thumb rhythmically.

"The one thing that amazes me," he said at length, "is that I have seen it—" He paused, running his fingers through his beard. "—that I have seen it . . . and I still live. What is it, Donald, and how has it found its way into your neat religious parlor?"

"I told you: it's a wild sphere. It's called Alexander's horned sphere. It's really not so extraordinary, Hans. There's a wild arc on the next page. Here's your brandy; now let's get started with the game." Don was impatient to forget the thing, for Hans to close the book and change the subject. He didn't like this reaction to a simple mathematical object, as though it were something more than it actually was.

"One could imagine Dali painting it," Hans continued, still fixing his gaze on the picture, "all ugly, bleeding lumps of flesh. That would be the obvious way. But I think I see it as a sculpture. It would have to be tremendous—say, thirty feet high—so that the branches start out fat as sequoias and end up—and end up microscopic—and never end. They should go on to the atomic level and beyond. Alexander's horned sphere. Can you see it squatting in the sun in the middle of Chicago, like some horrid, slimy crab? Yeah! Come on along, come on along. A sculpture, yes, a sculpture, that's the way I'd do it."

He slammed the book abruptly and laid it on the floor.

"Donald," he said, more in his natural voice. "Donald, if you have looked at that picture before and not felt the fear of the darkness in your veins, then all I can say is—you are hopelessly lost in your salvation. Have a Camel?"

Don shook his head, smiling, and Hans fitted a cigarette into his holder, continuing to talk between his teeth as he puffed life into it.

"Donald, the only man who is on such good terms with the devil that he can look him in the eye so casually is the satisfied theologian. You are a priest, not a prophet, and you must learn that even the bestest church what am will not protect you, your doctrinal orthodoxy will not save you, when the prophets begin to quake and wail outside the temple."

"And you are the prophet?" asked Don, egging him on.

"Hah! No, not *your* prophet. No, honey lamb, you've missed my point. Or else my analogy doesn't work out right. The prophets. Donald, I'm talking about mathematics, not art."

"Well, then I wish you wouldn't use a religious analogy. There is a fundamental difference in approach between religion and mathematics, and— No, let me finish. I know what you're going to say: that mathematics is predicated on the worship of reason. Well, that's wrong. Reason is only a tool to certain ends."

"Well, I agree, of course, Don. Reason is a tool to certain ends, and in your case those ends are basically theological. It's clear, you know, in this baroque fascination you have with the intricacies of your own proofs. You're only interested in plastering over the cracks in the temple. You've grown too dependent on it; you're afraid to worship in sunlight. Don't hide behind reason, Don. Your enemy will use the same tool. It's not reason that's against you, sweetheart; it's history."

"Anyway," Don interjected, annoyed at this turn in the conversation, "let's start playing or we'll never be done by ten."

"All right. But, Donald, I am going to do that sculpture, that Alexander's whatsit, someday. If ever I get a big commission. And you will be the only man who will understand all of its . . . all of its deeper meanings, my friend."

It was to be another fifteen years before he would get that commission and carry out his threat. It would be only twenty feet tall, not thirty, and in Cleveland, not Chicago. But it would shock, amaze, and frighten thousands of art lovers/haters, just as the original conception had shocked, amazed, and frightened Hans Kaefig.

Hans Kaefig, whose thoughts enveloped, surrounded, like rows of black disks, moving, shifting, unpredictably, while Donald Lucus's white disks coiled and struck, each move a step in a plan, each play a proverb. The patterns of black and white tesselations became too intricate to follow, and then there was no pattern at all, only the flashing black and white and a buzzing behind them, an insistent buzzing; as seconds stretched and expanded, he groped for his thoughts, sorting out the buzzing, reached for his glasses and the button on the intercom.

"Yes, Ruth?"

"It's two fifty-five, sir."

"Thank you, Ruth. Thank you."

He sat another minute, not really awake. Then, both hands on the chair, he lifted himself to his feet and did his best to tidy his suit. The Director was twenty years his junior, and yet he felt like a truant student being sent to the principal's office to explain himself, and knowing that the principal is never disposed to hear explanations.

3

He hadn't prepared a lecture in years. He had spoken of math only with other mathematicians. He had lost the knack of translation. There were English words, phrases, similes, that could say the same thing as a few swift logical statements, but he had forgotten them. And so he did not prepare a lecture. He had no idea what he was going to say to the Director, whether to present him with a neat, clear proof or simply to shout *"Gott ist tot,"* and make his point loudly and emotionally. He knew the Director was not a believer in any of the fundamental truths that were at stake. He would view a breakdown in the fabric of logic the same as a breakdown of the subway system. It was a nuisance to him, but it was certainly not his job to address himself to the problem; that was what metro engineers were paid for. That was what mathematicians were paid for. Lucus could not approach him on that level. What level he should direct his strategy toward he was not certain. He had, however, foreseen his opponent's moves well enough to expect the reception he received from the Director's private secretary.

"I'm sorry, Professor Lucus," the young man clipped, his glasses sliding down his nose in what seemed a studied parody of Ruth. "I'm sorry, but the Director is extremely busy this afternoon. I can make an appointment for Wednesday, I think . . . Of course, the Executive Board meeting is coming up next week . . ."

"That's all right, young man, I arranged to see him for a few minutes at three. I'll just slip in, and you can go back to your datebook." He would not be stalled any longer by the technical shunting about of the organization.

The Director looked blankly up from his desk as Lucus shut the door. He mumbled something feeble about thinking it was tomorrow that they were to meet, hoping to be rid of Lucus. As it became apparent that the math head had no intention of being put off further, he graciously conceded the skirmish and turned in an overly friendly manner to the problem itself.

As they talked, he leaned back in his chair, making full use of the physical advantage of his position. He sat comfortably in his shirtsleeves, collar open, bulky arms raised with his hands behind his head. Lucus alternately stood and sat—neither position was comfortable—in coat and tie, sweating through his shirt in the overheated office.

The Director was in his early forties, had held his position for three years. He divided his time unequally between his office, wife and children, and a girlfriend in San Jose. He drank more than Lucus had at his age, but seldom drank brandy. He was a mediocre chemist and an excellent administrator. He played golf one weekend out of two and worried that he was growing too fat. Lucus knew all this and very little else about the man. It was probable that he had studied calculus in college and forgotten a good deal of it by now, that he would be surprised to learn that an excellent mathematician might be very bad at arithmetic.

"Wait a minute," he protested before Lucus was very deep into his subject. "I thought Lobachevsky did that. I don't know much about math, but isn't that what non-Euclidean geometry is? Didn't they prove that Euclid was wrong? If it wasn't a big catastrophe then, why should it be now?"

And so he had to backtrack and try to give a ten-minute summary of the history of axiomatics. That Euclid's main contribution was not in his specific theorems, but in his method of assuming a very small number of "self-evident truths" and deriving all his results from them alone. That the question in the nineteenth century had only been over the notion of "self-evident," and then only over the fifth postulate, the so-called parallel postulate, and the exterior angle theorem. That non-Euclidean geometries had never denied the *consistency* of Euclid, but had only proposed alternative, equally consistent systems.

The Director balked at the word "consistency."

"But what's the difference between consistent and true?" he asked innocently.

"Truth has no meaning in mathematics," Lucus began. At the Director's scowl he corrected himself, for he was no logician, and these distinctions did not come quite naturally to him. "Or rather, truth is defined only relative to a given system of assumptions, you see. A statement is true in this system if it can be proved . . . I'm not sure if that's quite right . . . Well, anyway, if it necessarily follows from the assumptions. But a system of assumptions is consistent if you can't prove a contradiction from them, you see? If they could be a description of something that really exists."

"Okay, let me get this straight," said the Director, fishing a pack of Marlboros out of his pocket. "Something is *in*consistent if you can prove a contradiction from it, right? And what your Professor David seems to have done is prove that Euclid's postulates—is that the right word?—that his axioms or postulates or whatever are inconsistent. Am I right? So that means the whole notion of Euclidean geometry is nonsense. Well, I'm no mathematician, but I don't see the problem. Luckily this Russian has given you an alternative. So if, as you say, Euclid is scrapped, you still have this hyperbolic geometry and this other one, the one like the sphere, to choose from. It's very interesting, but hardly the kind of thing that requires any sort of executive decision."

Lucus bit his lip in frustration. He had always been a bad teacher, and Hans had said that . . .

"The non-Euclidean geometries were proven consistent by Riemann and Lobachevsky—" he began.

"Yes, well, that takes care of it, doesn't it?" the younger man interrupted.

"No, it doesn't!" said Lucus, too loudly. He sat down and tried to control his voice. "Non-Euclidean geometries were proven consistent by constructing models of them *within* Euclidean space. They are *conditionally* consistent. They are consistent only if Euclid is consistent. And, in the same way, Euclid depends on them. David's proof is valid for all three."

"You mean to say that *every* system of geometry is . . . is inconsistent . . . is meaningless?"

"Yes, sir. Not just geometry. Euclid can be derived from the real numbers. The real numbers can be derived from set theory. If Euclid is inconsistent, then the whole basis of mathematics is demolished. David's proof comprises the futility—" Donald Lucus' vision began to blur. His heart pumped blood deafeningly into his temples. There was a sharp pain in his chest. He spread his soaked and empty palms and spoke hoarsely. "—the futility of everything."

The Director was not unmoved by this display. He expected such an emotional plea on the part of a suppliant for a research grant on occasion. He was used to tearful outbursts from his girlfriend in San Jose, and he could react gently but unfeelingly in most emotional situations. But old men made him acutely uncomfortable. Emotional involvement in one's professional work puzzled and frightened him. He did not even yet understand the importance of the revelation which had been disclosed to him, but he did understand that it must be of some importance to bring this staid and dry old man to tears.

"You've checked it on the 666?" he asked.

Lucus looked away from him, embarrassed, fighting for breath, but trying not to breathe too deeply. "Not yet," he answered. "I have computer time tonight. I've made arrangements to have a social projection done this Friday, dependent on your approval."

"My approval?"

"For Limited Interest status."

"Oh." The Director rounded his lips meditatively and put his hands behind his head again. His cigarette lay in the ashtray, a long gray ash extending from the filter.

"Oh," he repeated. "Well, yes, of course. I suppose if it checks out, that you feel Something Must Be Done?"

"Yes, sir. I think there may be indications that Something Must Be Done about the problem."

And so Lucus knew that he had won the minimal confidence that he needed from the Director. The matter was to be given priority at the Executive Board meeting next week. He would have to go through the whole explanation again, many times. But it would be easier, the responsibility would no longer be entirely his. His white pieces coiled and struck across the board like a snake, squeezing the black ones out of strategic positions, reducing Hans's forces to a few holdouts near the edge. The brandy was sharp and exhilarating this evening.

4

With that trying interview over, Lucus felt a change in his mind and body. The oppressive burden was gone, and he could look forward to a great deal of time-and energy-consuming work. Responsibility was his, but it was the sort that he could be comfortable with, responsibility to get things done, to keep things moving. He spent the rest of Monday afternoon debugging the CONPROOF 2 input, which had arrived from the A50 during his absence. The work was routine, undemanding, and gently satisfying. By five-thirty it was in shape to run, and Lucus went home to dinner and eleven hours of cool and dreamless sleep.

Tuesday he worked continuously, stopping only for half an hour for lunch, fortified during the day by three cups of Ruth's dark but tasteless coffee. He had called Bibliography as soon as he got in and had his checklist headings augmented greatly. Every month articles containing certain key words or phrases in their titles or abstracts were sent to all the department heads. The controlling program was sophisticated enough to produce some very worthwhile information and very little that did not hold at least some interest for him. Now, in addition to his standard topology codes, he added a few checks in various kinds of geometry—it was, after all, likely to be his field of specialization into the foreseeable future.

David's proof had checked out perfectly in all situations, which did not surprise him at all. The program—which turned out to supply much more detailed output than he remembered from CONPROOF1—even made some suggestions on simplification of certain steps in the proof. It was indeed valid. There could be no doubt of that. He began writing up a short report on the program, which he would eventually include in his report to the Executive Board.

All day Wednesday was spent in conference with the head of the computer division, explaining in detail the results of David's proof and its connection with the rest of mathematics. They finally decided that the social projection could be done fairly easily with existing programs and data tapes, and it wouldn't be necessary to confer with Sociology—at least not until after the initial run. Lucus found himself working especially well with the man, developing an instant rapport and communicating the details of the problem much better than he had with the Director. In fact, the entire day he felt especially energetic and happy, almost euphoric, and he finally went home after seven with a genuine sense of accomplishment, disturbed only by the itching occasional thought that there was something he had meant to do but forgotten—nothing very important, but some detail that was left out, that destroyed the symmetry of the day. But this thought was eventually buried by the mass of other details, important details, enjoyable details, that competed for his time until late Wednesday evening.

Thursday and Friday were spent shuffling between two projects: the social projection and his report to the Executive Board. The Executive Board meeting, which had been scheduled to begin at two o'clock on Tuesday, was rescheduled to the morning, and all section heads were advised that some very important business might well cause it to run into most of the afternoon. The mere fact that the nature of this business was not mentioned, of course, tipped them off that it was "Limited Interest" and probably involved the sort of executive action that the Institute was theoretically not empowered to take. Actually, the Institute did stick literally to the guidelines in the Congressional bill which had authorized its founding. It served in an advisory capacity to the agencies which carried out the occasional difficult decisions which the board was sometimes forced to reach despite the seeming incompatibility of these decisions with the supposed concerns of the Institute. And if employees of the Institute were called upon to aid these other government agencies in the regrettable but necessary enforcement of decisions made in the interest of the general good, they clearly cooperated with their government as private individuals, usually in "special consultant" positions, and not as employees of the FBRI itself.

And so by adhering to the letter of its charter, the Institute managed to stretch the spirit of the charter when that spirit became a threat to more important considerations. No one on the Executive Board took this responsibility lightly. Indeed, it was the gravity, the solemnity with which they were bound to weigh questions of ethics and then exercise their own benign power in the interests of the whole of society—it was this gravity, the awesome weight of obligation and the crucial necessity of judicious application of their superior skills which secretly thrilled many of the board members and added an unequaled zest to these meetings. There had been one department head who opposed all such actions, but he had left the Institute to return to teaching some years back. Now there was usually a broad and healthy range of opinion and discussion on questions of "interference," as it was called, Genetics holding out against exercise of such power except in the most extreme cases, Biophysics being perhaps a bit overzealous in his enthusiasm for the Institute's potential control of future events, and the rest of the departments arranging themselves variously between these two poles as befitted their individual politics, esthetics, professional ethics, temperaments, and digestive difficulties.

Mathematics, that is to say, Dr. Donald Lucus, was never entirely sure where he belonged in the spectrum, being, he knew, too easily swayed by each side of the debate in its turn, and most often casting his ballot with the majority. The issues were always too vague, uncontrollable, and, as he put it, "political," and they seemed very far from his real concerns in his work. This time, however, he had no doubt which side of the issue he would take. He would have to hold the floor himself, and he knew for certain that the Institute must take appropriate measures to head off the catastrophic events that could be instigated by David's proof.

The initial surprise of the board when they realized it was he who was going to read the report was the customary reaction he got from everybody whenever they learned that Mathematics might be involved in something important. He was the out man in the building, and he now felt a bit of pride in presenting his case, in being allowed to overshadow their scientific concerns with the problems of *his* field for an entire morning, perhaps for a whole workday.

He began by giving a precise and ordered account of David's proof, its connectioin with the previous history of mathematics, and the interrelation of geometry with the foundations of all modern mathematical theory.

After this, and before the presentation by Computers and Sociology of the social projection, there was a period of questions directed to Lucus, as was customary, to ensure that everyone had a clear understanding of the issue. As it was, a number of them did, but the others were often reluctant to question a speaker in another field, less for fear of exposing their own ignorance than out of professional courtesy and a desire to avoid any question which might appear as a challenge to the speaker's competence. It was usually understood that each head considered his colleagues as the final experts in their own fields, and their private terrain must be respected, as they respected his. An attempt to gain too complete an understanding of his territory was dangerously close to a take-over of his sovereign province. So the questions were only halfhearted requests for clarification about Hilbert's axioms and the independence of the parallel postulate. Until Genetics raised a fluttering hand and asked with a Socratic smirk, "You say that the principles of Euclidean geometry can be derived from set theory, and, of course, this can be verified on the 666?"

"That's right."

"And so it all falls back on the principles of symbolic logic, and theoretically you could put the whole thing in terms of a proof in logic?"

"Yes, in fact, there is a program which can do just that with most mathematical proofs, and once I get around to it I intend—"

"Yes, yes. Well, very good. But isn't it true that your whole method of proof is based on symbolic logic?"

"Yes."

"And since you use this same method of proof in David's theorem—"

Lucus smiled. He knew what was being suggested.

"Since you use this same method of proof in David's theorem, and since you have *shown* that this method of proof is not valid—I mean, that's what you've shown by proving the inconsistency of symbolic logic itself—then you really haven't proved David's theorem at all, have you? I mean, have you?"

Naturally, most of the board members were made acutely uncomfortable by this want of tact, and especially by Genetics' toothy grin as she spread her hands and waited for an answer. She was most unpopular, and it was rumored that she was not likely to remain much longer at the Institute. These rumors, however, had been circulated for a number of years with no noticeable effect on her position or her unwillingness to initiate a change of career herself. She was fifty-five now, and if she remained at the Institute many more years she was quite likely to become the Grand Old Lady, in which case her position and power would be unchallengeable until her own gracious retirement.

But now she was a minority of one, smiling at Lucus that same wide friendly smile that her secretary had conscientiously striven to imitate, smiling and waiting. He was not prepared for this particular question, but he had taught basic math courses long enough to be familiar with its general tactics. It was, put on the grossest level, to dismiss mathematical jargon as a lot of nonsense. But, applied more subtly, it consisted of pointing out illogicality in the nature of the mathematical approach, in the detection of ubiquitous paradoxes, all of which eventually boiled down to some variation on the Russell paradox: the serpent of mathematics was forever swallowing its own tail. But Russell had long ago found a solution in the simple expedient of multiplying the number of serpents and lining them up to swallow each other's tails, a much more plausible situation, and happily one which introduced no further problems until the level of transfinite numbers was reached, and here the mathematician was again swimming in his own medium.

"But your argument can only serve to confirm David's proof. You are arguing from a paradox, from the absurd." Lucus returned her smile, feeling around his shoulders the temporary, illusory mantle of the Grand Old Man.

"How so?" she asked.

"Well," he continued, "simply because the method of proof of David's theorem is invalid, that does not insubstantiate its result—no, wait, let me finish—at best you would have to conclude that it is proven neither true nor false. Now suppose you assume that the basis of logic is in fact *valid*. Then you are forced to accept David's theorem and the proof of the *in*validity of your logic; you are led to a contradiction. Therefore the assumption of the validity of logic is untenable."

"But, on the other hand," she objected, "if you assume that logic is invalid, then David's proof is invalid."

"Precisely. And that is a perfectly consistent position. David's theorem does *not* say 'This sentence is false.' It says 'This sentence is unprovable,' and therefore it must be true."

"I think you're talking in circles," said Genetics.

"It's all very well for you to think that, but the fact remains that this position is sound—it can easily be verified on the 666."

The remaining questions were dutiful inquiries into the nature of the Russell paradox, and each answer was followed by a polite "Oh, yes, I see." Lucus was calm and confident by the time Computers and Sociology began their description of the social projection.

The problem had been of quite a different nature from that of most of the projections that had been introduced to the board. Usually, a discontinuity was introduced into a percentage prediction pattern and the other initial conditions were varied within certain ranges, so that the effects of the invention of some new device or some new discovery in physical law could be ascertained, both short-and long-range effects. The discontinuity had some direct and immediate effect on material or political conditions, on arms capabilities or projected population figures or the economy. A good deal of such research had been done, it was true, with discontinuities of a religious or philosophical nature, and the sort of results obtained was quite well understood by those in the field. The Institute as a whole was seldom concerned with such results. It was the responsibility of other branches of the government to deal with the possible detrimental long-range effects of new religious or philosophical movements.

The projection for David's theorem showed remarkably unperturbed figures for a long period. Even in the mathematical world, it was predicted, little notice would be taken of it for fifteen to twenty years, notwithstanding its immediate effect on Dr. Lucus. It would be dismissed and ignored—at first. But within thirty years the disruption of the mathematical world would become violent and begin spreading into other fields. Still it would remain an academic debate. There would be much name-calling and side-taking, the introduction of heated emotions into decisions of hiring, tenure, structuring of mathematics and science departments. But still the public at large would remain entirely unaware of the issue. New schools of philosophy would arise to address themselves to the problem. Within forty years the issue would be taken up in the public press, the result being an increase in the polarization of scientists in all fields, and within fifty years, sixty at the outside, a violent antiscience reaction at all levels of government, huge cutbacks in funding, reduction of departments in universities all over the country, massive shutdowns of laboratories, and even elimination of many industrial research programs. The original issue would be mostly forgotten by this time; it would be the widespread fear of being dominated by scientists, scientists pictured as caricatures from Vincent Price movies, that would be the main concern of the public. But the effects would be disastrous for the scientific community.

There was heated discussion on this projection until one o'clock, when the board took an hour break, and then reconvened for the afternoon session. There was particular objection to the long-range nature of the projection. Many members of the board felt it was not their duty to be concerned with developments half a century in the future, and some of them were highly skeptical of the accuracy of the 666's figures for such a period, although error estimates were given for all figures, and they usually didn't get beyond 5 or 10 percent at the sixty-year level.

To many it still seemed incredible that a mathematical theorem could have such an effect, but the evidence of the 666 was hard to dispute. Only Genetics objected to the input and assumptions of the projection program itself.

"What you don't assume," she said to Lucus, again smiling with all her teeth, "is the ingenuity of the world's mathematicians. The program projects the proof of one theorem into the future, without considering what else will be proved in the future. If you scrap geometry, why shouldn't a new geometry arise? If you scrap Aristotelian logic, why shouldn't a new logic arise?"

"I assure you, madam," answered Lucus, "the proof is valid for *all* logics, for intuitionism as well as for the many-valued systems."

"And there are no other roads?" she asked cynically.

"And there are no other roads."

"Well, then, I would suppose your field is likely to come to a stagnant standstill in any case, with or without any help from Professor David."

There were shocked murmurs of censure up and down the table. Cryogenics, who had intended to question the methods of formulation of the theorem for the 666, thought better of it and folded his hands in silence.

The debate went on, but there was a marked increase of support for Lucus. By seven, when the final ballot was taken, he sat content, watching the other men and women folding their slips of paper. There was a sense of camaraderie, of common purpose, that he had never felt before with the Institute personnel. It was much like serving on the Honor Code Committee in college: the shared duty, the secret debates, and the final pride in the satisfying justice of the verdict, a day very well spent creeping, creeping to a close.

It was the consensus of the Executive Board that Something Must Be Done, and with that most of them could entirely forget the problem in good conscience. For Lucus, however, there were more meetings with the Director and Sociology, conferences in Washington, and an eventual temporary advisory post in an agency of the executive branch.

He returned to his work at the Institute, allotting a few hours a week to his new advisory position, the few hours he used to devote to research, research which seemed to lose some of its insistent appeal now that he had other important duties. The fall *Quarterly* came out only two weeks late, minus David's article, with a brief editorial apology for the delay. Letters had to be written, editors conferred with to assure the difficulty of David's publishing his article elsewhere in the near future. Meanwhile, the foundation was laid for the discrediting of any publication he eventually managed to achieve. The strategy was all laid out by experts in Washington who had handled similar cases, and Lucus had only to implement a few of the moves which required the prestige of his position in his field. Fortunately, David was up for tenure at his university that year. When it was not granted, he had immense difficulty obtaining a position elsewhere. He was quite a meek man, and certainly not paranoid enough to accuse anyone of being involved in a conspiracy against him. He ended up turning to high school teaching, which allowed him less and less time for any serious research.

The whole problem was neatly and efficiently disposed of, and Lucus could not help admiring the simplicity of the plan of action. He had carried out his own part of the program carefully and professionally. No one could reproach him. He had represented his profession admirably.

5

"You're a creep!" said Hans, coming up behind him on his way back to the dorm after class.

"What do you mean by that?" he asked without turning, gripping his books tighter, feeling the strength in his fingers against them.

"You're a creep, Lucus, that's what!" Hans held up the morning's edition of the school paper. "Look at this, you creepy bastard! That was really a rotten thing to do."

"Look, the whole committee voted on Jonathan!" Don left the brick walk and cut across the lawn, anxious to be rid of Kaefig.

"And I know how you voted too, you creep!" Hans insisted. "And now the poor kid'll be expelled, just because he got caught cribbing on one little exam . . . and you can be smug about it. Can't you find a better way to save the honor of your precious code?"

"It wasn't one little exam, Hans, and besides—let go of me!—and besides, it wasn't just Jonathan we had to think of; it was the integrity of the whole school and the honor code. How long do you—I said let go of me!—how long do you think the faculty would let us keep the honor code if we let everybody off who broke it? Hey, get away! I've got to go to lunch!"

He dropped all his books as Hans wrestled him to the ground. It wasn't much of a match; Don was a skinny kid, and he was more worried about keeping his glasses from slipping off than in putting up a fight. He was soon on his back with Hans on top. Hans swiftly pulled off his glasses and slapped him twice, hard. He lay still, his face stinging and wet, as Hans got up.

"I had to do that, Donnie, because of the rotten thing you did to Jonathan. You understand that, don't you? I mean, we don't have to talk about it anymore. That was all I wanted to say." He held up the palm of his right hand. "You okay now?"

Don nodded. His face was hot with fear, his eyes turned upward, away from Hans.

"See you tonight for a beer?"

Don nodded. He closed his eyes and dug the fingers of one hand into the cool soil. But Hans still waited, out of his field of vision, tired and afraid; unsure, Hans waited for an answer.

"Kaefig, you're crazy!" he whispered.

He lay there, not moving, long after Hans had gone. The tree above was blurred, but he could trace the pattern of its larger branches, and the smaller ones seemed to wink in and out of existence. If he concentrated and squinted, he could follow even those, tracing the patterns over and over with his eyes and his mind. He often looked at trees, never tired of looking at trees, sliding along the limbs with his eyes, absorbing the whole of the latticework.

Every year there were new branches on the tree outside his window. Even this year, strange as it might seem, new green branches on the tree outside the window of the elephants' graveyard. To lose himself on these branches. To reach up and out with his mind. To lie prone on earth and cease to ponder on himself, the while he stared at nothing, drawn nowhere. His breath came with more difficulty these days, his hands would not close with ease and pained him when they did. Walking was an effort and all chairs too hard and wrongly proportioned. He longed to—what was the verse?—to

seek release
From dusty bondage into luminous air.

But he was no hero. What lay on his desk was merely dry and inevitable. The morning's mail, a cup of coffee, the laughing icosahedron. He had thumbed through the journals marked for his attention by Bibliography late in the morning. And there it was, in a Polish journal of logic, to be sure. In German, yes, but there could be no mistake: *"Die Widerspruchlichkeit der Logikgrundsätze als Folge eines geometrischen Beweises,"* by Kálmán Kodály of the University of Budapest.

Of course, it had to happen; anyone should have known. He placed one hand on the icosahedron, no longer needing to look at it, and raised himself to his feet. He walked slowly to the window, knowing he would find something there, the vision of order and neutrality, of "light anatomized." It was waiting for him, soft and green and easy on his eyes. And for long minutes that morning, Donald Lucus stood at his window, tracing the lines, the beautiful lines of the tree, and wept for the death of his dear friend Hans.

MARC LAIDLAW

Love Comes to the Middleman

Upon the wall, the neighborlings were arguing. Jack listened to the piping voices with increasing anger. The problems of the little people sounded all too much like his own, except smaller.

He opened his eyes and searched for the offending home among the array of tiny buildings stacked to the ceiling of his room. In most, the lights were dim or out completely; in a few, tiny shadows moved against the curtains. The smell of almond tobacco smoke drifted from half-open doorways; newspapers rustled. As a rule, the smaller citizens went to sleep early, and those who stayed up kept their voices down once he'd turned off his light.

Tonight, the Pewlins were the noisemakers:

"If you can't stay inside your budget, pretty soon we won't have a budget!"

"It's not me wasting money on drink and gambling."

"It's not you making money, either. I need my recreation."

"Recreation? You're a drunk with bad luck. It's not like you're developing a skill. You just get drunker and unluckier. And the next time—"

On his knees now, Jack rapped sharply on the door of the Pewlins' house with a fingernail. "Hey, in there. I've got a heavy day tomorrow."

At the sound of his voice, curtains stirred in the windows of other houses. The Pewlins, too embarrassed to face him, merely began to mutter.

"Told you you'd wake him. We're going to lose this house and end up in somebody's sock drawer."

"Oh, shut up. I'm going to bed."

As Jack crawled back into his bed—a lumpy mattress laid out on the floor—someone scratched on his door. With a sigh, he got up and opened it.

His house was halfway up the wall of the next room. The giant and his wife shared that room. She was out there now, leaning so close to his door that he could have stepped onto her nose.

"Having trouble in there, Jack?

"It's the Pewlins again. They went to bed. Thanks for asking, Nairla."

"If they're any trouble, we'd be glad to take them out here. We can hardly hear them, they're so small. I know the neighborlings' voices can be so penetrating when you're trying to sleep."

How do you know that? he thought, but didn't ask. He had kept her awake a few times, no doubt, with his infrequent parties.

"No, seriously, it's not a problem now. Thanks anyway." He leaned out of the doorway and she turned her head so that he could whisper into her vast ear: "I think you've probably intimidated them."

She pulled back and smiled, a very nice smile. Nairla had always taken a special interest in him; for his part, he'd always been attracted to big-boned red-headed women. But not as big as Nairla. She was quite out of his league. And besides, her enormous husband lay out there like a range of hills, snoring away. The houses and office buildings along the giants' walls were all dark; Jack's samesize neighbors kept similar hours. He only wished his neighborlings could be so quiet.

"Sleep tight," Nairla said.

"Would you ask her to keep it down?" piped a voice from a corner of Jack's room. "Some of us have to get up in the morning."

Jack awoke with a groan on his lips and a vile taste in his mouth, and the complaints of the neighborlings in his ears: "Turn off that alarm clock! We're awake!"

As he reached out to switch off the alarm, he realized that he was sick. Swimming head, upset stomach—the flu had been going around at the office. This had to be it. He would just lie here a while and hope it didn't get worse.

False hope. He lurched out of bed and ran into the bathroom. When he looked up from the sink, the houses along the window ledge were coming to life. Complaints came drifting down to him: "Was that birdsong I heard? What a way to wake up."

"Sorry," Jack said.

From bed, he called the office. The phones weren't being answered yet. He would have to lie and wait a while.

An hour later, he awoke to the sound of buzzing. Tiny private fliers darted among the buildings on his wall. Some of them maneuvered around the ceiling of the room, caught in elaborate flight patterns as they waited their turns to exit through the vents near the ceiling, then headed for neighborling pueblos in other houses. The configurations confused him; they were like specks swimming across his eyes.

Late, he thought. I'm late.

He sat up abruptly and grabbed the phone, fighting nausea as he dialed the office. Mrs. Clorn sounded mildly amused by his illness; apparently she didn't believe him. As he hung up, a tiny voice asked, "Jack, are you sick?"

He glanced up at the nearest wall. A young mother and her child stood on the ramp outside their house.

"Oh, Revlyn, hi. Yeah, I've got the flu."

"Wish I could help. I make soup for Tilly when he's sick . . . but you know how much I'd have to make for you."

"Don't worry, I'll be fine. It's a twenty-four-hour bug."

Someone scratched on his door.

"Come in!" he called.

Nairla put her eye to the opening. "I thought you were still in there. Aren't you going to work today?"

"He's got the flu," Revlyn called.

"What's that?" asked Nairla. "Is she talking to me, Jack?"

"Of course I'm talking to you, you dumb giant! Can't you hear me?" Revlyn broke off into wild laughter.

"Nice to see you too, dear," Nairla said.

"She says I'm sick," said Jack. "I've got the flu."

"The flu? Oh dear! Would you like some oatmeal? I can mash it up for you. Plenty of fluids and what else? Do you have a fever? I'd lend you a thermometer but . . . you know."

"Why don't you ask her if any of your neighbors are home," Revlyn said. "You need someone to look in on you, Jack. Seriously."

"What's that?" Nairla said.

"Nothing," Jack said.

"Ask her, Jack," said Revlyn. "You'd be silly not to. What if you get really sick?"

He sighed. "Nairla, Revlyn says I should ask you to see if any of my neighbors are home, in case I get worse."

"That's a good idea. I'll check. There are a few . . . oh, I have a wonderful idea! I've been meaning to introduce you two for the longest time. She's an artist."

Oh no, he thought. Not that.

"Nairla—" he began. But she was gone. He looked up at Revlyn. "She's going to get somebody."

"Good. That's nice of her. I'm sorry I make fun of her, but she is so deaf, you know?"

"She's big, that's all." Big and nosy.

A minute passed, in which he heard Nairla humming to herself, vibrating the walls of his house. He thought he was going to be sick again. Footsteps came up along the ramp, then a face peeked around the door. It was a samesize woman, a redhead with big bones, strong hands. As she came all the way into the room, she said, "You must be Jack. You look pretty sick."

He wondered how long Nairla had been waiting for this chance.

"I'm Liss. I brought some tea. I was just making up a pot for my morning work." She sat on a corner of the mattress and poured some tea into a water glass he'd left on the floor.

"So, uh, I hear you're an artist," he said.

She handed him the glass. "I'm a sculptor. Mostly I apply for grants."

"You're an artist and you get paid for it?"

"The Plenary Council—have you heard of them? Everything I do goes to the Council, and they arrange showings. There's a lot of interest in us, among the giants. And I'm talking about *giant* giants—bigger than Nairla. They're intrigued by our perceptions of the world. Do you realize they have to look at our art under microscopes?"

"Art, huh? So what's it mean to the little guy?"

She reached in her pocket and took out both a magnifying glass and a little box. "I'll show you. This was made by a sculptor three sizes smaller than us. He's been a great influence on my own work, though I can't say I'm nearly as good as him. The detail work is incredible."

Liss handed him first the glass, and then the box with the lid taken off. He found himself staring into a construction the size of a rice grain, elaborately carved, a piece of microscopic scrimshaw showing spiral staircases that grew smaller and smaller as they curled toward the center of the grain. On the stairs were incredibly lifelike figures, also dwindling as the steps shrank. Looking at it made him dizzy. He thought of himself looking at the tiny stairs, and of Nairla looking in at him, and of someone looking in at Nairla.

He blinked at Liss. "Do . . . do the giants have art?"

"Sure. It's hard for us to see it sometimes, though. You have to get way back. We've tried scaling it down through the levels, but it loses something. The size is part of the meaning."

"That's really interesting."

"Do you think so? It's funny, with all of us living on this wall, I spend more time talking to Nairla and the neighborlings than I ever do with people my own size."

He shrugged. "I'm like that. I have a boring job; it makes me feel like all the samesizes are boring. When I get home, I don't want to see anybody my own size."

"My husband's the same way," Revlyn called. "I can't get him to take me out. He'd rather stay home and watch the little people."

Jack held out the glass and Liss refilled it. He smiled at her, feeling better already, and raised it in a toast.

"Here's to a new friend," he said. He was gratified to see her blush.

At the doorway, Nairla blinked in and said, "Aren't you two cute?"

"Oh, spare me," said Revlyn, and went inside.

ROBERT SHECKLEY

Miss Mouse and the Fourth Dimension

I first met Charles Foster at the Claerston Award dinner at Leadbeater's Hall in the Strand. It was my second night in London. I had come to England with the hope of signing some new authors for my list. I am Max Seidel, publisher of Manjusri Books. We are a small, esoteric publishing company operating out of Linwood, New Jersey—just me and Miss Thompson, my assistant. My books sell well to the small but faithful portion of the population interested in spiritualism, out-of-body experiences, Atlantis, flying saucers, and New Age technology. Charles Foster was one of the men I had come to meet.

Pam Devore, our British sales representative, pointed Foster out to me. I saw a tall, good-looking man in his middle thirties, with a great mane of reddish-blond hair, talking animatedly with two dowager types. Sitting beside him, listening intently, was a small woman in her late twenties with neat, plain features and fine chestnut hair.

"Is that his wife?" I asked.

Pam laughed. "Goodness, no! Charles is too fond of women to actually marry one. That's Miss Mouse."

"Is 'Mouse' an English name?"

"It's just Charles's nickname for her. Actually, she's not very mouselike at all. Marmoset might be more like it, or even wolverine. She's Mimi Royce, a society photographer. She's quite well off—the Royce textile mills in Lancashire, you know—and she adores Charles, poor thing."

"He does seem to be an attractive man," I said.

"I suppose so," Pam said, "if you like the type." She glanced at me to see how I was taking that, then laughed when she saw my expression.

"Yes, I *am* rather prejudiced," she confessed. "Charles used to be rather interested in me until he found his own true love."

"Who was—?"

"Himself, of course. Come, let me introduce you."

Foster knew about Manjusri Books and was interested in publishing with us. He thought we might be a good showcase for his talents, especially since Paracelsus Press had done so poorly with his last, *Journey Through the Eye of the Tiger*. There was something open and boyish about Foster. He spoke in a high, clear English voice that conjured up in me a vision of punting on the Thames on a misty autumn day.

Charles was the sort of esoteric writer who goes out and has adventures and then writes them up in a portentous style. His search was for—well, what shall I call it? The Beyond? The Occult? The Interface? Twenty years in this business and I still don't know how to describe, in one simple phrase, the sort of book I publish. Charles Foster's last book had dealt with three months he had spent with a Baluchistani dervish in the desert of Kush under incredibly austere conditions. What had he gotten out of it? A direct though fleeting knowledge of the indivisible oneness of things, a sense of the mystery and grandeur of existence. . . . In short, the usual thing. And he had gotten a book out of it; and that, too, is the usual thing.

We set up a lunch for the next day. I rented a car and drove to Charles's house in Oxfordshire. It was a beautiful old thatch-roofed building set in the middle of five acres of rolling countryside. It was called Sepoy Cottage, despite the fact that it had five bedrooms and three parlors. It didn't actually belong to Charles, as he told me immediately. It belonged to Mimi Royce.

"But she lets me use it whenever I like," he said. "Mouse is such a dear." He smiled like a well-bred child talking about his favorite aunt. "She's so interested in one's little adventures, one's trips along the interface between reality and the ineffable. . . . Insists on typing up my manuscripts just for the pleasure it gives her to read them first."

"That is lucky," I said, "typing rates being what they are these days."

Just then Mimi came in with tea. Foster regarded her with bland indifference. Either he was unaware of her obvious adoration of him, or he chose not to acknowledge it. Mimi, for her part, did not seem to mind. I assumed that I was seeing a display of the British National Style in affairs of the heart—subdued, muffled, unobtrusive. She went away after serving us, and Charles and I talked auras and ley lines for a while, then got down to the topic of real interest to us both—his next book.

"It's going to be a bit unusual," he told me, leaning back and templing his fingers.

"Another spiritual adventure?" I asked. "What will it be about?"

"Guess!" he said.

"Let's see. Are you by any chance going to Machu Picchu to check out the recent reports of spaceship landings?"

He shook his head. "Elton Travis is already covering it for Mystic Revelations Press. No, my next adventure will take place right here in Sepoy Cottage."

"Have you discovered a ghost or poltergeist here?"

"Nothing so mundane."

"Then I really have no idea," I told him.

"What I propose," Foster said, "is to create an opening into the unknown right here in Sepoy Cottage, and to journey through it into the unimaginable. And then, of course, to write up what I've found there."

"Indeed," I said.

"Are you familiar with Von Helmholtz's work?"

"Was he the one who read tarot cards for Frederick the Great?"

"No, that was Manfried Von Helmholtz. I am referring to Wilhelm, a famous mathematician and scientist in the nineteenth century. He maintained that it was theoretically possible to *see* directly into the fourth dimension."

I turned the concept over in my mind. It didn't do much for me.

"This 'fourth dimension' to which he refers," Foster went on, "is synonymous with the spiritual or aethereal realm of the mystics. The name of the place changes with the times, but the region itself is unchanging."

I nodded. Despite myself, I am a believer. That's what brought me into this line of work. But I also know that illusion and self-deception are the rule in these matters rather than the exception.

"But this spirit realm or fourth dimension," Foster went on, "is also our everyday reality. Spirits surround us. They move through that strange realm which Von Helmholtz called the fourth dimension. Normally they can't be seen."

It sounded to me like Foster was extemporizing the first chapter of his book. Still, I didn't interrupt.

"Our eyes are blinded by everyday reality. But there are techniques by means of which we can train ourselves to see what *else* is there. Do you know about Hinton's cubes? Hinton is mentioned by Martin Gardner in *Mathematical Carnival*. Charles Howard Hinton was an eccentric American mathematician who, around 1910, came up with a scheme for learning how to visualize a tesseract, also called a hypercube or four-dimensional square. His technique involved colored cubes which fit together to form a single master cube. Hinton felt that one could learn to see the separate colored cubes in the mind, and then, mentally, to manipulate and rotate them, fold them into and out of the greater cube shape, and to do this faster and faster until at last a gestalt forms and the hypercube springs forth miraculously in your mind."

He paused. "Hinton said that it was a hell of a lot of work. And later investigators, according to Gardner, have warned of psychic dangers even in attempting something like this."

"It sounds like it could drive you crazy," I said.

"Some of those investigators *did* wig," he admitted cheerfully. "But that might have been from frustration. Hinton's procedure demands an inhuman power of concentration. Only a master of yoga could be expected to possess that."

"Such as yourself?"

"My dear fellow, I can barely remember what I've just read in the newspaper. Luckily, concentration is not the only path into the unknown. Fascination can more easily lead us to the mystic path. Hinton's principle is sound, but it needs to be combined with Aquarian Age technology to make it work. That is what I have done."

He led me into the next room. There, on a low table, was what I took at first to be a piece of modernistic sculpture. It had a base of cast iron. A central shaft came up through its middle, and on top of the shaft was a sphere about the size of a human head. Radiating in all directions from the sphere were Lucite rods. At the end of each rod was a cube. The whole contraption looked like a cubist porcupine with blocks stuck to the end of his spines.

Then I saw that the blocks had images or signs painted on their faces. There were Sanskrit, Hebrew, and Arabic letters, Freemason and Egyptian symbols, Chinese ideograms, and other figures from many different lores. Now the thing no longer looked to me like a porcupine. Now it looked like a bristling phalanx of mysticism, marching forth to do battle against common sense. And even though I'm in the business, it made me shudder.

"He didn't know it, of course," Foster said, "but what Hinton stumbled upon was the mandala principle. His cubes were the parts; put them all together in your mind and you create the Eternal, the Unchanging, the Solid Mandala, or four-dimensional space, depending upon which terminology you prefer. Hinton's cubes were a three-dimensional exploded view of an aethereal object. This object refuses to come together in our everyday reality. It is the unicorn who flees from the view of man—"

"—but lays its head in the lap of a virgin," I finished for him.

He shrugged it off. "Never mind the figures of speech, old boy. Mouse will unscramble my metaphors when she types up the manuscript. The point is, I can use Hinton's brilliant discovery of the exploded mandala whose closure produces the ineffable object of endless fascination. I can journey down the endless spiral into the unknown. This is how the trip begins."

He pushed a switch on the base of the contraption. The sphere began to revolve, the Lucite arms turned, and the cubes on the ends of those arms turned, too, creating an effect both hypnotic and disturbing. I was glad when Foster turned it off.

"My Mandala Machine!" he cried triumphantly. "What do you think?"

"I think you could get your head into a lot of trouble with that device," I told him.

"No, no," he said irritably. "I mean, what do you think of it all as the subject for a book?"

No matter what else he was, Foster was a genuine writer. A genuine writer is a person who will descend voluntarily into the flaming pits of hell for all eternity, as long as he's allowed to record his impressions and send them back to earth for publication. I thought about the book that would most likely result from Foster's project. I estimated its audience at about one hundred and fifty people including friends and relatives. Nevertheless, I heard myself saying, "I'll buy it." That's how I manage to stay a small and unsuccessful publisher despite being so smart.

I returned to London shortly after that. Next day I drove to Glastonbury to spend a few days with Claude Upshank, owner of the Great White Brotherhood Press. We have been good friends, Claude and I, ever since we met ten years ago at a flying saucer convention in Barcelona.

"I don't like it," Claude said, when I told him about Foster's project. "The mandala principle is potentially dangerous. You can really get into trouble when you start setting up autonomous feedback loops in your brain like that."

Claude had studied acupuncture and Rolfing at the Hardrada Institute in Malibu, so I figured he knew what he was talking about. Nevertheless, I thought that Charles had a lot of savvy in these matters and could take care of himself.

When I telephoned Foster two days later, he told me that the project was going very well. He had added several refinements to the Mandala Machine: "Sound effects, for one. I'm using a special tape of Tibetan horns and gongs. The overtones, sufficiently amplified, can send you into instant trance." And he had also bought a strobe light to flash into his eyes at six to ten beats a second: "The epileptic rate, you know. It's ideal for loosening up your head." He claimed that all of this deepened his state of trance and increased the clarity of the revolving cubes. "I'm very near to success now, you know."

I thought he sounded tired and close to hysteria. I begged him to take a rest.

"Nonsense," he said. "Show must go on, eh?"

A day later, Foster reported that he was right on the brink of the final breakthrough. His voice wavered, and I could hear him panting and wheezing between words. "I'll admit it's been more difficult than I had expected. But now I'm being assisted by a certain substance which I had the foresight to bring with me. I am not supposed to mention it over the telephone in view of the law of the land and the ever-present possibility of snoops on the line, so I'll just remind you of Arthur Machen's 'Novel of the White Powder' and let you work out the rest for yourself. Call me tomorrow. The fourth dimension is finally coming together."

The next day Mimi answered the telephone and said that Foster was refusing to take any calls. She reported him as saying that he was right on the verge of success and could not be interrupted. He asked his friends to be patient with him during this difficult period.

The next day it was the same, Mimi answering, Foster refusing to speak to us. That night I conferred with Claude and Pam.

We were in Pam's smart Chelsea apartment. We sat together in the bay window drinking tea and watching the traffic pour down the King's Road into Sloane Square. Claude asked, "Does Foster have any family?"

"None in England," Pam said. "His mother and brother are on holiday in Bali."

"Any close friends?"

"Mouse, of course," Pam said.

We looked at each other. An odd presentiment had occurred to us simultaneously, a feeling that something was going terribly wrong.

"But this is ridiculous," I said. "Mimi absolutely adores him, and she's a very competent woman. What could there be to worry about?"

"Let's call once more," Claude said.

We tried, and were told that Mimi's telephone was out of order. We decided to go to Sepoy Cottage at once.

Claude drove us out in his old Morgan. Mimi met us at the door. She looked thoroughly exhausted, yet there was a serenity about her which I found just a little uncanny.

"I'm so glad you've come," she said, leading us inside. "You have no idea how frightening it's all been. Charles came close to losing his mind in these last days."

"But why didn't you tell us?" I demanded.

"Charles implored me not to. He told me—and I believed him—that he and I had to see this thing through together. He thought it would be dangerous to his sanity to bring in anyone else at this point."

Claude made a noise that sounded like a snort. "Well, what happened?"

"It all went very well at first," Mimi said. "Charles began to spend increasingly longer periods in front of the machine, and he came to enjoy the experience. Soon I could get him away only to eat, and grudgingly at that. Then he gave up food altogether. After a while he no longer needed the machine. He could see the cubes and their faces in his head, could move them around at any speed he wanted, bring them together or spread them apart. The final creation, however, the coming together of the hypercube, was still eluding him. He went back to the machine, running it now at its highest speed."

Mimi sighed. "Of course, he pushed himself too hard. This time, when he turned off the machine, the mandala continued to grow and mutate in his head. Each cube had taken on hallucinatory solidity. He said the symbols gave off a hellish light that hurt his eyes. He couldn't stop those cubes from thundering through his mind. He felt that he was being suffocated in a mass of alien signs. He grew agitated, swinging quickly between elation and despair. It was during one of his elated swings that he ripped out the telephone."

"You should have sent for us!" Claude said.

"There was simply no time. Charles knew what was happening to him. He said we had to set up a counter-conditioning program immediately. It involved changing the symbols on the cube faces. The idea was to break up the obsessive image trains through the altered sequence. I set it up, but it didn't seem to work for Charles. He was fading away before my eyes, occasionally rousing himself to murmur, 'The horror, the horror . . .'"

"Bloody hell!" Claude exploded. "And then?"

"I felt that I had to act immediately. Charles's system of counter-conditioning had failed. I decided that he needed a different sort of symbol to look at—something simple and direct, something reassuring—"

Just then Charles came slowly down the stairs. He had lost a lot of weight since I had seen him last, and his face was haggard. He looked thin, happy, and not quite sane.

"I was just napping," he said. "I've got rather a lot of sleep to catch up on. Did Mouse tell you how she saved what little is left of my sanity?" He put his arm around her shoulders. "She's marvelous, isn't she? And to think that I only realized yesterday that I loved her. We're getting married next week, and you're all invited."

Mimi said, "I thought we were flying down to Monte Carlo and getting married in the city hall."

"Why, so we are." Charles looked bewildered for a moment. He touched his head with the unconscious pathos of the wounded soldier in the movie who hasn't yet realized that half his head is blown away. "The old think-piece hasn't quite recovered yet from the beating I gave it with those wretched cubes. If Mimi hadn't been here, I don't know what would have happened to me."

They beamed at us, the instant happy couple produced by Hinton's devilish cubes. The transformation of Charles's feelings toward Mimi—from fond indifference to blind infatuation—struck me as bizarre and dreamlike. They were Svengali and Trilby with the sexes reversed, a case of witchcraft rather than love's magic.

"It's going to be all right now, Charles," Mimi said.

"Yes, love, I know it is." Charles smiled, but the animation had gone out of his face. He lifted his hand to his head again, and his knees began to sag. Mimi, her arm around his waist, half supported and half dragged him to the stairs.

"I'll just get him up to bed," she said.

Claude, Pam, and I stood in the middle of the room, looking at each other. Then, with a single accord, we turned and went into the parlor where the Mandala Machine was kept.

We approached it with awe, for it was a modern version of ancient witchcraft. I could imagine Charles sitting in front of the thing, its arms revolving, the cubes turning and flashing, setting up a single ineradicable image in his mind. The ancient Hebrew, Chinese, and Egyptian letters were gone. All of the faces of all the cubes now bore a single symbol—direct and reassuring, just as Mimi had said, but hardly simple. There were twenty cubes, with six faces to a cube, and pasted to each surface was a photograph of Mimi Royce.

ISAAC ASIMOV

The Feeling of Power

Jehan Shuman was used to dealing with the men in authority on long-embattled Earth. He was only a civilian but he originated programming patterns that resulted in self-directing war computers of the highest sort. Generals consequently listened to him. Heads of congressional committees too.

There was one of each in the special lounge of New Pentagon. General Weider was space-burned and had a small mouth puckered almost into a cipher. Congressman Brant was smooth-cheeked and clear-eyed. He smoked Denebian tobacco with the air of one whose patriotism was so notorious, he could be allowed such liberties.

Shuman, tall, distinguished, and Programmer-first-class, faced them fearlessly.

He said, "This, gentlemen, is Myron Aub."

"The one with the unusual gift that you discovered quite by accident," said Congressman Brant placidly. "Ah." He inspected the little man with the egg-bald head with amiable curiosity.

The little man, in return, twisted the fingers of his hands anxiously. He had never been near such great men before. He was only an aging low-grade technician who had long ago failed all tests designed to smoke out the gifted ones among mankind and had settled into the rut of unskilled labor. There was just this hobby of his that the great Programmer had found out about and was now making such a frightening fuss over.

General Weider said, "I find this atmosphere of mystery childish."

"You won't in a moment," said Shuman. "This is not something we can leak to the firstcomer. Aub!" There was something imperative about his manner of biting off that one-syllable name, but then he was a great Programmer speaking to a mere technician. "Aub! How much is nine times seven?"

Aub hesitated a moment. His pale eyes glimmered with a feeble anxiety. "Sixty-three," he said.

Congressman Brant lifted his eyebrows. "Is that right?"

"Check it for yourself, Congressman."

The congressman took out his pocket computer, nudged the milled edges twice, looked at its face as it lay there in the palm of his hand, and put it back. He said, "Is this the gift you brought us here to demonstrate. An illusionist?"

"More than that, sir. Aub has memorized a few operations and with them he computes on paper."

"A paper computer?" said the general. He looked pained.

"No, sir," said Shuman patiently. "Not a paper computer. Simply a sheet of paper. General, would you be so kind as to suggest a number?"

"Seventeen," said the general.

"And you, Congressman?"

"Twenty-three."

"Good! Aub, multiply those numbers, and please show the gentlemen your manner of doing it."

"Yes, Programmer," said Aub, ducking his head. He fished a small pad out of one shirt pocket and an artist's hairline stylus out of the other. His forehead corrugated as he made painstaking marks on the paper.

General Weider interrupted him sharply. "Let's see that."

Aub passed him the paper, and Weider said, "Well, it looks like the figure seventeen."

Congressman Brant nodded and said, "So it does, but I suppose anyone can copy figures off a computer. I think I could make a passable seventeen myself, even without practice."

"If you will let Aub continue, gentlemen," said Shuman without heat.

Aub continued, his hand trembling a little. Finally he said in a low voice, "The answer is three hundred and ninety-one."

Congressman Brant took out his computer a second time and flicked it. "By Godfrey, so it is. How did he guess?"

"No guess, Congressman," said Shuman. "He computed that result. He did it on this sheet of paper."

"Humbug," said the general impatiently. "A computer is one thing and marks on paper are another."

"Explain, Aub," said Shuman.

"Yes, Programmer. Well, gentlemen, I write down seventeen, and just underneath it I write twenty-three. Next I say to myself: seven times three—"

The congressman interrupted smoothly, "Now, Aub, the problem is seventeen times twenty-three."

"Yes, I know," said the little technician earnestly, "but I *start* by saying seven times three because that's the way it works. Now seven times three is twenty-one."

"And how do you know that?" asked the congressman.

"I just remember it. It's always twenty-one on the computer. I've checked it any number of times."

"That doesn't mean it always will be, though, does it?" said the congressman.

"Maybe not," stammered Aub. "I'm not a mathematician. But I always get the right answers, you see."

"Go on."

"Seven times three is twenty-one, so I write down twenty-one. Then one times three is three, so I write down a three under the two of twenty-one."

"Why under the two?" asked Congressman Brant at once.

"Because—" Aub looked helplessly at his superior for support. "It's difficult to explain."

Shuman said, "If you will accept his work for the moment, we can leave the details for the mathematicians."

Brant subsided.

Aub said, "Three plus two makes five, you see, so the twenty-one becomes a fifty-one. Now you let that go for a while and start fresh. You multiply seven and two, that's fourteen, and one and two, that's two. Put them down like this and it adds up to thirty-four. Now if you put the thirty-four under the fifty-one this way and add them, you get three hundred and ninety-one, and that's the answer."

There was an instant's silence and then General Weider said, "I don't believe it. He goes through this rigmarole and makes up numbers and multiplies and adds them this way and that, but I don't believe it. It's too complicated to be anything but hornswoggling."

"Oh no, sir," said Aub in a sweat. "It only *seems* complicated because you're not used to it. Actually the rules are quite simple and will work for any numbers."

"Any numbers, eh?" said the general. "Come, then." He took out his own computer (a severely styled GI model) and struck it at random. "Make a five seven three eight on the paper. That's five thousand seven hundred and thirty-eight."

"Yes, sir," said Aub, taking a new sheet of paper.

"Now"—more punching of his computer—"seven two three nine. Seven thousand two hundred and thirty-nine."

"Yes, sir."

"And now multiply those two."

"It will take some time," quavered Aub.

"Take the time," said the general.

"Go ahead, Aub," said Shuman crisply.

Aub set to work, bending low. He took another sheet of paper and another. The general took out his watch finally and stared at it. "Are you through with your magic-making, Technician?"

"I'm almost done, sir. Here it is, sir. Forty-one million, five hundred and thirty-seven thousand, three hundred and eighty-two." He showed the scrawled figures of the result.

General Weider smiled bitterly. He pushed the multiplication contact on his computer and let the numbers whirl to a halt. And then he stared and said in a surprised squeak, "Great Galaxy, the fella's right."

The President of the Terrestrial Federation had grown haggard in office and, in private, he allowed a look of settled melancholy to appear on his sensitive features. The Denebian War, after its early start of vast movement and great popularity, had trickled down into a sordid matter of maneuver and countermaneuver, with discontent rising steadily on Earth. Possibly, it was rising on Deneb too.

And now Congressman Brant, head of the important Committee on Military Appropriations, was cheerfully and smoothly spending his half-hour appointment spouting nonsense.

"Computing without a computer," said the president impatiently, "is a contradiction in terms."

"Computing," said the congressman, "is only a system for handling data. A machine might do it, or the human brain might. Let me give you an example." And, using the new skills he had learned, he worked out sums and products until the president, despite himself, grew interested.

"Does this always work?"

"Every time, Mr. President. It is foolproof."

"Is it hard to learn?"

"It took me a week to get the real hang of it. I think you would do better."

"Well," said the president, considering, "it's an interesting parlor game, but what is the use of it?"

"What is the use of a newborn baby, Mr. President? At the moment there is no use, but don't you see that this points the way toward liberation from the machine. Consider, Mr. President"—the congressman rose and his deep voice automatically took on some of the cadences he used in public debate—"that the Denebian War is a war of computer against computer. Their computers forge an impenetrable shield of countermissiles against our missiles, and ours forge one against theirs. If we advance the efficiency of our computers, so do they theirs, and for five years a precarious and profitless balance has existed.

"Now we have in our hands a method for going beyond the computed, leapfrogging it, passing through it. We will combine the mechanics of computation with human thought; we will have the equivalent of intelligent computers, billions of them. I can't predict what the consequences will be in detail, but they will be incalculable. And if Deneb beats us to the punch, they may be unimaginably catastrophic."

The president said, troubled, "What would you have me do?"

"Put the power of the administration behind the establishment of a secret project on human computation. Call it Project Number, if you like. I can vouch for my committee, but I will need the administration behind me."

"But how far can human computation go?"

"There is no limit. According to Programmer Shuman, who first introduced me to this discovery—"

"I've heard of Shuman, of course."

"Yes. Well, Dr. Shuman tells me that in theory there is nothing the computer can do that the human mind cannot do. The computer merely takes a finite amount of data and performs a finite number of operations upon them. The human mind can duplicate the process."

The president considered that. He said, "If Shuman says this, I am inclined to believe him—in theory. But, in practice, how can anyone know how a computer works?"

Brant laughed genially. "Well, Mr. President, I asked the same question. It seems that at one time computers were designed directly by human beings. Those were simple computers, of course, this being before the time of the rational use of computers to design more advanced computers had been established."

"Yes, yes. Go on."

"Technician Aub apparently had, as his hobby, the reconstruction of some of these ancient devices, and in so doing he studied the details of their workings and found he could imitate them. The multiplication I just performed for you is an imitation of the workings of a computer."

"Amazing!"

The congressman coughed gently. "If I may make another point, Mr. President—the further we can develop this thing, the more we can divert our federal effort from computer production and computer maintenance. As the human brain takes over, more of our energy can be directed into peacetime pursuits and the impingement of war on the ordinary man will be less. This will be most advantageous for the party in power, of course."

"Ah," said the president, "I see your point. Well, sit down, Congressman, sit down. I want some time to think about this. But meanwhile, show me that multiplication trick again. Let's see if I can't catch the point of it."

Programmer Shuman did not try to hurry matters. Loesser was conservative, very conservative, and liked to deal with computers as his father and grandfather had. Still, he controlled the West European computer combine, and if he could be persuaded to join Project Number in full enthusiasm, a great deal would be accomplished.

But Loesser was holding back. He said, "I'm not sure I like the idea of relaxing our hold on computers. The human mind is a capricious thing. The computer will give the same answer to the same problem each time. What guarantee have we that the human mind will do the same?"

"The human mind, Computer Loesser, only manipulates facts. It doesn't matter whether the human mind or a machine does it. They are just tools."

"Yes, yes. I've gone over your ingenious demonstration that the mind can duplicate the computer, but it seems to me a little in the air. I'll grant the theory, but what reason have we for thinking that theory can be converted to practice?"

"I think we have reason, sir. After all, computers have not always existed. The cave men with their triremes, stone axes, and railroads had no computers."

"And possibly they did not compute."

"You know better than that. Even the building of a railroad or a ziggurat called for some computing, and that must have been without computers as we know them."

"Do you suggest they computed in the fashion you demonstrate?"

"Probably not. After all, this method—we call it 'graphitics,' by the way, from the old European word 'grapho,' meaning 'to write'—is developed from the computers themselves, so it cannot have antedated them. Still, the cave men must have had *some* method, eh?"

"Lost arts! If you're going to talk about lost arts—"

"No, no. I'm not a lost art enthusiast, though I don't say there may not be some. After all, man was eating grain before hydroponics, and if the primitives ate grain, they must have grown it in soil. What else could they have done?"

"I don't know, but I'll believe in soil growing when I see someone grow grain in soil. And I'll believe in making fire by rubbing two pieces of flint together when I see that too."

Shuman grew placative. "Well, let's stick to graphitics. It's just part of the process of etherealization. Transportation by means of bulky contrivances is giving way to direct mass transference. Communications devices become less massive and more efficient constantly. For that matter, compare your pocket computer with the massive jobs of a thousand years ago. Why not, then, the last step of doing away with computers altogether? Come, sir, Project Number is a going concern; progress is already headlong. But we want your help. If patriotism doesn't move you, consider the intellectual adventure involved."

Loesser said skeptically, "What progress? What can you do beyond multiplication? Can you integrate a transcendental function?"

"In time, sir. In time. In the last month I have learned to handle division. I can determine, and correctly, integral quotients and decimal quotients."

"Decimal quotients? To how many places?"

Programmer Shuman tried to keep his tone casual. "Any number!"

Loesser's lower jaw dropped. "Without a computer?"

"Set me a problem."

"Divide twenty-seven by thirteen. Take it to six places."

Five minutes later Shuman said, "Two point oh seven six nine two three."

Loesser checked it. "Well, now, that's amazing. Multiplication didn't impress me too much because it involved integers, after all, and I thought trick manipulation might do it. But decimals—"

"And that is not all. There is a new development that is, so far, top secret and which, strictly speaking, I ought not to mention. Still—we may have made a breakthrough on the square root front."

"Square roots?"

"It involves some tricky points and we haven't licked the bugs yet, but Technician Aub, the man who invented the science and who has an amazing intuition in connection with it, maintains he has the problem almost solved. And he is only a technician. A man like yourself, a trained and talented mathematician, ought to have no difficulty."

"Square roots," muttered Loesser, attracted.

"Cube roots, too. Are you with us?"

Loesser's hand thrust out suddenly. "Count me in."

General Weider stumped his way back and forth at the head of the room and addressed his listeners after the fashion of a savage teacher facing a group of recalcitrant students. It made no difference to the general that they were the civilian scientists heading Project Number. The general was the overall head, and he so considered himself at every waking moment.

He said, "Now square roots are all fine. I can't do them myself and I don't understand the methods, but they're fine. Still, the project will not be sidetracked into what some of you call the fundamentals. You can play with graphitics any way you want to after the war is over, but right now we have specific and very practical problems to solve."

In a far corner Technician Aub listened with painful attention. He was no longer a technician, of course, having been relieved of his duties and assigned to the project, with a fine-sounding title and good pay. But, of course, the social distinction remained, and the highly placed scientific leaders could never bring themselves to admit him to their ranks on a footing of equality. Nor, to do Aub justice, did he, himself, wish it. He was as uncomfortable with them as they with him.

The general was saying, "Our goal is a simple one, gentlemen—the replacement of the computer. A ship that can navigate space without a computer on board can be constructed in one-fifth the time and at one-tenth the expense of a computer-laden ship. We could build fleets five times, ten times, as great as Deneb could if we could but eliminate the computer.

"And I see something even beyond this. It may be fantastic now, a mere dream, but in the future I see the manned missile!"

There was an instant murmur from the audience.

The general drove on. "At the present time our chief bottleneck is the fact that missiles are limited in intelligence. The computer controlling them can only be so large, and for that reason they can meet the changing nature of antimissile defenses in an unsatisfactory way. Few missiles, if any, accomplish their goal, and missile warfare is coming to a dead end, for the enemy, fortunately, as well as for ourselves.

"On the other hand, a missile with a man or two within, controlling flight by graphitics, would be lighter, more mobile, more intelligent. It would give us a lead that might well mean the margin of victory. Besides which, gentlemen, the exigencies of war compel us to remember one thing. A man is much more dispensable than a computer. Manned missiles could be launched in numbers and under circumstances that no good general would care to undertake as far as computer-directed missiles are concerned . . ."

He said much more, but Technician Aub did not wait.

Technician Aub, in the privacy of his quarters, labored long over the note he was leaving behind. It read finally as follows:

"When I began the study of what is now called graphitics, it was no more than a hobby. I saw no more in it than an interesting amusement, an exercise of mind.

"When Project Number began, I thought that others were wiser than I, that graphitics might be put to practical use as a benefit to mankind, to aid in the production of really practical mass-transference devices perhaps. But now I see it is to be used only for death and destruction.

"I cannot face the responsibility involved in having invented graphitics."

He then deliberately turned the focus of a protein depolarizer on himself and fell instantly and painlessly dead.

They stood over the grave of the little technician while tribute was paid to the greatness of his discovery.

Programmer Shuman bowed his head along with the rest of them but remained unmoved. The technician had done his share and was no longer needed, after all. He might have started graphitics, but now that it had started, it would carry on by itself overwhelmingly, triumphantly, until manned missiles were possible with who knew what else.

Nine times seven, thought Shuman with deep satisfaction, is sixty-three, and I don't need a computer to tell me so. The computer is in my own head.

And it was amazing the feeling of power that gave him.

HENRY H. GROSS ______.

Cubeworld

I · CRISIS

"Geometry Rules."

This angular axiom, flanked by straightedge and compass and inscribed in aluminum above the arched entrance to the Planned Planethood Institute in upper New York State, passed briefly over my head as I was ushered inside by a pair of armed and uniformed federal agents.

They had been polite as butlers and as unyielding as boulders when they'd rousted me from bed at the state university campus, where I was chairman of the Communications Department and, some liked to think, internationally recognized as something of a media expert. All they would tell me was that my country, indeed my world, needed my immediate services—along with those of several dozen other specialists. I must come at once. No, this was not a volunteer position. Thank you, we'll wait.

I dressed, told Sarah, my wife, not to worry—a statement for which I had no basis whatsoever—and kissed our daughter, Rebecca, gently on her cheek as she slept. Then I followed my escorts into their long dark vehicle and let myself be transported some fifteen miles to the rustic headquarters of Planned Planethood—an organization that had been in existence for less than two weeks and which was yet a virgin to the greedy glare of publicity. This was in fact precisely the state of affairs I had been called upon to reverse. Indeed, as I was to learn, the very survival of our species depended on all 5.1 billion of us knowing about Planned Planethood and its objectives within a matter of weeks—and then following its directives to a T. Or at least up to humanity's theoretical "cooperation threshold," which Planned Planethood had already calculated to be 79.36 percent.

I was led down a corridor, up in an elevator, and then along yet another corridor to a door marked "Operations Room." There I was greeted by a large man with a dynamic smile and curly gray hair who I recognized immediately as the renowned and respected Mexican statesman, Rufus Cortez, a figure as famous for his global political machinations as for his humble insistence that folks of all stations call him Rufus. He shook my hand and invited me into his large, ad hoc headquarters, in which were assembled some forty-odd men, women, and teenagers of varying nationalities, staring at me as if their proceedings had been pending solely upon my arrival. Rufus gave me a ruddy smile, dismissed my keepers with a wave, and introduced me to the crowd. "Most of you have heard of William Lindsay, who herewith completes our primary team," he said briefly. "Besides teaching, he's chief media consultant for the United Nations, as well as public relations advisor for Iceland, Venezuela, and my own country, Mexico. I know him as well by Secret Service dossier. I'm sure you will find Bill as trustworthy and superior as any of yourselves, who of course received similar scrutiny before being selected for this crucial international endeavor. Naturally, we have assembled the world's best for this most awesome of all tasks facing us." To me, he added, "Welcome, *amigo*."

"Thank you," I said, still not knowing why I'd been invited—if invited was the word—to this particular party. I ran my fingers along my unshaved cheek and waited for Rufus to continue.

"Our planet's very survival is at stake," he said and pointed to a large chart on which had been rendered a sketch labeled *The Salton Mass*. "As you can see for yourself."

I looked at the drawing, recognizing our solar system but drawing a blank on the nearby arrow with the crosshatching on it. "Can't quite say as I do," I admitted.

Gesturing to a pretty young lady who blushed and lowered her eyes, Rufus elaborated, "We've named it, naturally, after Maureen Salton, who discovered it just a couple of weeks ago using a new computer algorithm she invented. Maureen is an astronomy student doing graduate work at Harvard and something of a prodigy, as well."

I nodded politely to the young woman.

"Apparently," Rufus went on, "a massive cloud of superdense particles is surging toward our solar system in a trajectory that

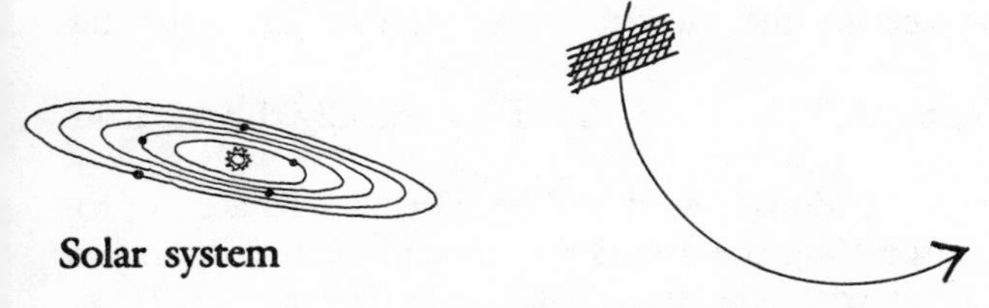

Figure 1 The Salton Mass

threatens to profoundly affect the gravitational equilibrium of the entire system before it continues off into the universe at large. Ms. Salton has calculated, and others have since confirmed, that unless we take drastic action promptly, this gravitational tsunami will cause the orbits of our system's planets to wobble and change direction. This will cause calamitous results on Earth, including climatic and meteorological disruptions, agricultural chaos, and possible orbital collisions, for example, with material in the asteroid belt.

I frowned, thinking of my family. "Sounds like a hell of a mess," I said.

"Getting hit with a rock the size of Florida would indeed be a mess," he agreed. "However," he added cheerfully, "we don't intend to let our planet slide down the cosmic tube without one hell of a fight. To that end, red tape has been cut to shreds in every government in the world in order to assemble Planned Planethood. By unanimous agreement, I have been appointed Chief of Operations, with unlimited emergency powers to meet the crisis. In addition to Maureen, who will be functioning as Planned Planethood's Chief of Astronomy, we have gathered an international team of experts in many other fields, including seismology, agriculture, oceanography, management, economy, transportation, demolitions, military affairs, and so forth. You, Bill, are to be our Chief of Communications, responsible for the crucial task of coordinating the minds of over five billion people and seeing to it that everyone knows what to do—and will *want* to—when the time comes to do it. Equally important will be rumor control. IBM has already initiated emergency production of over five billion inexpensive voiceprint sensors that can be attached to any radio or TV set. Thus, people can be certain the words they're hearing are verifiably yours."

"The horse's mouth," I quipped, drawing a friendly chuckle from the group.

Rufus beamed. "Additionally, it will be your task to act as project historian, chronicling the events that take place and issuing an official report on it when the crisis will finally have passed."

"But what can be done?" I asked, feeling uncomfortably adrift in a vast sea of IQ.

"A plan has already been devised," said Rufus. "It is to construct in space a very massive chunk of matter in the shape of a cube with provisions so that the cube's center of gravity and attitude in space can readily be altered, as desired, by the redirection of large quantities of water amongst various of the cube's six faces. In this way, we can compensate for the gravitational effects of the Salton Mass, in effect countering one imbalance with another. Now, why a cube, you ask?"

"All right. Why a cube?"

"Because it has corners. Ears, if you will. If we used a sphere, of which shape there is no shortage in the heavens, it couldn't get a gravitational 'purchase' on the rest of the solar system. It would 'spin its wheels' like a car stuck in mud but achieve no result. At the other extreme, a long, massive cylinder, for example, or a slab, would be *too* asymmetrical—too difficult, too slow, and requiring too much energy to maneuver quickly." Rufus smiled in that charismatic way of his that had melted the hearts of millions. "A cube, therefore, turns out to be the ideal compromise."

I felt as awed as a two-year-old. "A square planet?" I managed to say.

"Cubic," said Rufus.

"What are you going to call it?"

"Cubeworld."

"Makes sense."

"Si, amigo."

"But isn't something like that going to take a heck of a lot of material? What are you going to make it out of?"

"Earth," he said.

"You mean dirt? Mud?"

He shook his head, while the rest of his elite crew stared at me intently. "Earth," he repeated. "The."

II · TRANSFORMATION

The contemplated feat of engineering was, of course, colossal. In theory, we had to lop sections off some parts of the Earth (sorry 'bout that, Europe!) and redistribute the material elsewhere, specifically into eight three-sided corners each equidistant from its neighboring corners (see Figure 2), as if cubing a grapefruit by slicing six domes off its surface and building up "ears" out of the excavated material.

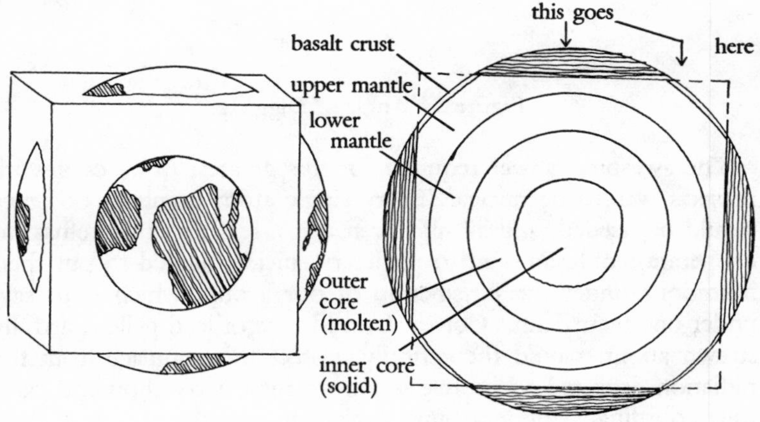

Figure 2 The Challenge

To our perception (our Sphereworld perception, that is, since Cubeworld perception, as you know, turned out to be markedly different) we would be digging "down" to begin with at an angle of fifty-three degrees, an angle which, in bizarre disregard of the fact that we'd be proceeding unvaryingly along an absolutely straight line, would nonetheless gradually change until it became ninety degrees and we were digging "straight ahead" instead of "down" (see Figure 3). The material thus chiseled away would simultaneously be piled up behind us at the same angle as our digging, resulting in a geometrically flat plane extending in all directions until intersection with neighboring planes.

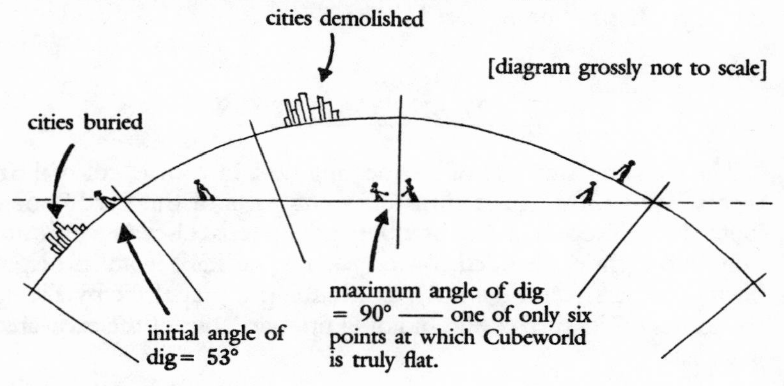

Figure 3 Angle of Digging

The awesome power required for this greatest of all earthwork projects was to be nuclear. Every single atom bomb in existence would be needed, and all nuclear nations, realizing that refusal to cooperate would amount to planetary suicide, agreed to put their munitions under the jurisdiction of our group, which is to say, under one man: Rufus Cortez. Untold tons of lead pellets, artfully strewn about, would theoretically protect the populace from the radiation expected to be released from these sixty thousand carefully coordinated underground explosions.

Although the total renovation of the entire globe promised to be no mean task, on the other hand humanity was not without resources. Our tabulations revealed the existence of some 200,000 airplanes, twenty million dump trucks, 6 million bulldozers and comparable numbers of like machinery. China alone was slated to contribute almost a billion wheelbarrows, not to mention much of the muscle power to wield them.

My job, of course, was to smooth the way, peoplewise, for all the high-tech engineering to come. When, as many of you will recall, I finally broke the news to the world, my voiceprint oozed pure optimism. In addition to setting forth the facts, laced with reassurances that our unprecedented construction scheme would be radioactively safe, I conveyed Rufus Cortez's personal promise that not one person would be left behind when the great demolition derby began. (In fact, over fifty thousand *chose* to remain behind and died.)

Preparations for the day on which the actual transformation would begin, designated Cube Day, covered a period of several months. The Earth was surveyed. Equipment was positioned. Armies were placed on alert. Workers with jackhammers scored the planet like glass in six great circles so that it would break off cleanly when tapped by our nuclear warheads. Although the logistics of all this were as massive as the mass itself, happily a sense of global comradeship took hold, as peoples of all nations put aside their differences and began exhibiting a degree of cooperation measured at 72.44 percent, commendably close to the theoretical limit of 79.36 percent I mentioned earlier. Because the transformation was expected to take about a month to complete, 5.1 billion thirty-day survival kits were prepared and distributed, and plans were coordinated for providing food, water, clothing, shelter, sewage disposal, and health care, including new baby deliveries, as needed. My own efforts during this period were directed toward helping people adjust to the great dislocations to come, preparing them to relinquish psychologically things they might hold dear: for example, the Empire State Building, the Mississippi River, Italy. Because there existed the great unknown of whether humanity could adapt to life on a cube or, indeed, a cube itself to partnership in the solar system, a worldwide campaign was conducted, under my supervision, to select humanity's most treasured artifacts—books, films, machinery, seeds of all kinds, frozen sperm and eggs, and so forth—and send as much of this material as our rocket capacity would allow out of the solar system altogether, so that if humanity did perish in its bold effort to save its world, someday our "bottles" on the ocean of space might be found and our civilization remembered, or possibly even revived.

A week before Cube Day people relocation began. My wife (a homebody), my daughter (a born explorer), and I were housed at Planned Planethood headquarters. There was a savory sense of tension amongst the populace, not unlike that before an important football game.

Finally, just hours before the beginning of the transformation, the globally televised firing of humanity's legacy into space took place from Cape Canaveral and other launching sites. A hundred rockets, with a combined payload of over half a million tons, blazed into the sky. Our ark was launched—or so we had been craftily led to believe. Unsuspecting, we breathed a collective sigh of relief; we might die now, but our important papers, as it were, were safe.

That afternoon, however, as I made a final tour with Rufus Cortez and our Russian Chief of Demolitions, Vladimir Dubrov, I noticed, with anxiety, that the vast apparatus for rapidly covering the earth with a carpet of lead pellets didn't seem to be operational; no one was at his or her designated post, as planned.

"You observe correctly, Mr. Lindsay," said the Chief. "We're not going to use nuclear explosives at all. Never were, in fact, although I deliberately didn't let this be known to anyone, including you. *Especially* you."

I stared at him in bewilderment.

"Instead," he went on, "we're going to slice our domed sections off the earth with an oil exploration technique called 'geothermal fracturing'—pumping water into hot rock to crack it. As the rock crumbles, we'll cart it away and fracture the next layer, continuing until we have achieved the shape we want. We figure the planet's going to be one big humid steambath for three to four weeks. That's why we provided everyone with rain slickers."

"Then why did you have me tell everyone we were going nuclear?" I asked, mildly annoyed at my wasted efforts and possibly sullied reputation, yet somehow fundamentally relieved.

"I had to say that to get the countries that had nuclear weapons to turn them over. Otherwise, they never would have." Rufus' cheeks bulged with a smile. "You saw that king-sized space launch this morning, didn't you? The one that supposedly shot our 'time capsules' into outer space?"

"I covered it, as you well know."

His black eyes twinkled. "Like a cub reporter, *amigo*—never double-checking the payloads to see if maybe the cargos had been switched. Some Chief of Communications!"

"You mean—?"

His grin widened. "Gone forever," he said to the sky. "All thirty thousand warheads, defused and heading harmlessly out to interstellar space. The Sword of Damocles lifted at last!"

I smiled too. "Why, you crafty old wizard. I thought you were given this job because you could be trusted!"

"*Si, si.* And, as you can see, I can be." He gave his aide a signal. "Here we go, *amigo*. Hang on to your seat belt!"

With that, the transformation began. Forget continental drift; this was the continental runs. With steam restricting vision to a few feet, much of the work was done by feel and straightedge. Bulldozers roared and swarmed. Landmarks disappeared forever as earth was piled upon them, cooled, and then used as foundation for further extension. Geography as we knew it ceased to exist. Billions slept on cots, perspiring in the humidity. Thunderstorms reigned. Oceans relocated violently, producing phenomena for whose underwater details we must await future dolphinic exposition. Computers tirelessly updated our logistics. Populations were uprooted on an hour's notice. The ground trembled, volcanos blasted out angrily, edges crawled toward apexes.

And thirty days later, in the late spring of 1987, the task was done, and humanity was left with nothing but a square, bare polyhedron and a smattering of food, water, supplies, and equipment. Gradually, the steam disappeared, the clouds wept their last, and we saw our first sunshine in a month. From Planned Planethood headquarters, Sarah, Rebecca, and myself gazed for the first time upon our new world. And as we did, we experienced in full measure a feeling that had been growing every day during the transformation: that we were somehow off balance, tilted, as if we were standing on a ramp. Oddly enough, when the world had *been* round, it had *felt* flat. Now that it truly *was* flat, it felt as if we were living in a concavity whose apparent grade ranged from flat to a forty-five-degree incline depending on where between the center and the corners you were standing at the moment of plumb-bob measurement. The four apexes visible from our side were like mountains looming thousands of miles high, poking, indeed, right out of the atmosphere altogether, which had remained essentially spherical, though not quite to the extent it had been before.

So, while our surveyors jubilantly certified our world euclideanly planer, it *felt* as though we were about to slide along the inside of a

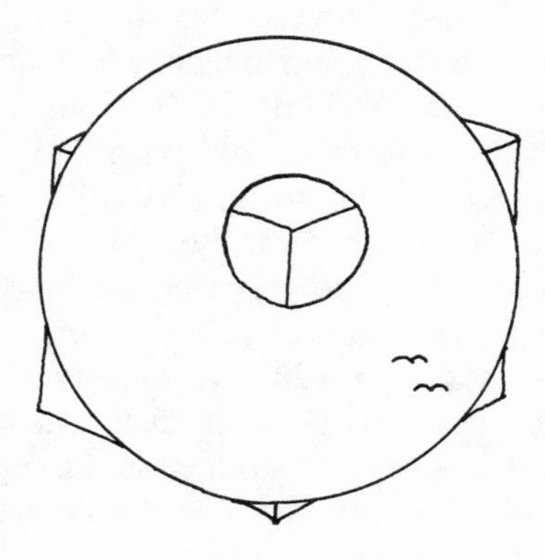

Figure 4 The Atmosphere

bowl toward its center, where, needless to say, the ocean had long since flowed and settled, and above which it now hovered, a jiggling bubble several miles high and 3,500 miles in diameter, one of six such bodies of water. Because Cubeworld *felt* steep, within minutes our planet *looked* steep as well. Though our horizon *was* straight, we *felt* it as bowed, and soon we began to "see" it that way, too (see Figure 5), in the same way as psychology test subjects who wear special inverting lenses readjust mentally and "see" the image right side up.

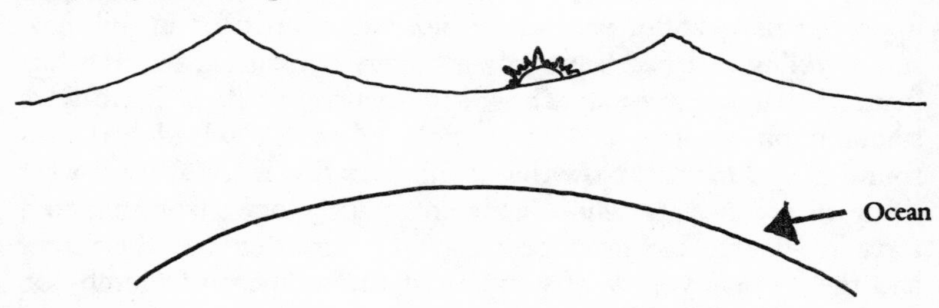

Figure 5 Inhabitants' Eye View of Cubeworld

So, what was Cubeworld *really,* I remember reflecting, as the sun slipped oddly over the crescent horizon of our recast planet, shrouding our "valley" in a lovely violet twilight? Flat or curved? Was "is-ness" what you felt, what you saw, or what you thought?

When senses tug in contradictory directions, which is to be believed? Or was instead the lesson Cubeworld taught us that truth is not to be found in *any* of the senses, and that perhaps, in the end, the only truth is shape? As Planned Planethood's slogan suggested: Geometry Rules. Is-ness is Angle.

I looked at Sarah, holding Rebecca close to her and looking out, or rather "up," across the vastness of our new land. She was crying.

III · CUBEWORLD

That first year was Cubeworld's honeymoon. After an initial period of inconvenience, for which I had psychologically readied everyone, resettlement grew geometrically more routine; and while some of the properties of Cubeworld took some getting used to, by and large most people found our revised planet a distinct improvement over the old one.

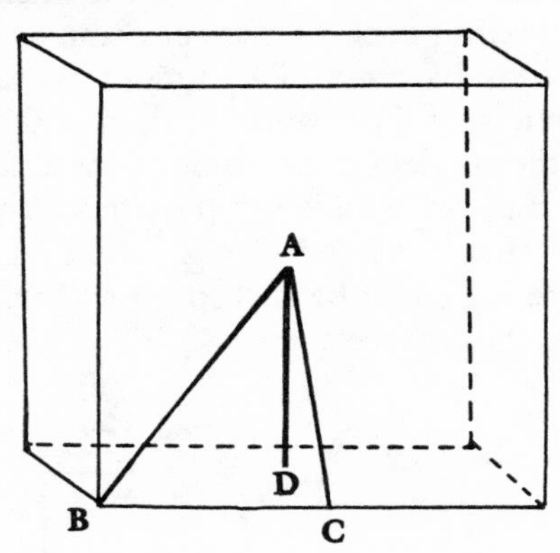

Figure 6 Cubeworld's Gravity

The new shape of gravity was, of course, the fundamental difference (see Figure 6). On a perfect sphere, all points on the surface are equidistant from its center; indeed, this is what defines the object *as* a sphere. On a cube, however, there is no single equidistant radius from center to surface but an infinite number of lengths bounded by three unique extremities: the ***shortest distance (AD),*** between the cube's center and one of the six centers of the six sides; the ***longest distance (AB),*** between the center and one of the eight corners; and an ***intermediate distance (AC),*** between the cube's center and the center of one of its twelve edges. All other distances fit somewhere within these three.

As a result, the uniform pull of gravity toward the center of our new world was intercepted by flat planes at different angles, depending on where on the plane you were standing. Let me put it to you straight (a word itself of increasingly ambiguous meaning): Cubeworld was one relentless incline. From apex *B* you'd roll to *C* or *D;* from *C* you'd roll to *D;* and at *D* you'd be six to ten miles underwater owing to the fact that an ocean had rolled there long before you did.

In its capacity as emergency gyroscope to the solar system, Cubeworld had been designed to finesse these immense blobs of water from one side to another at will, thereby continuously countering the effects of the Salton Cycle when that insidious cloud of inertia finally made the first of its two scheduled passes near our solar merry-go-round. This capability was realized within three months by connecting each of the six oceans to its four nearest neighboring oceans by pairs of enormous buried pipes that ran "up" to the centers of Cubeworld's edges and then "down" the other side into the adjacent ocean, though these conduits were generally thought of as "in" and "out" (not "hot" and "cold"; let's be serious here!). Thusly herded along by hundreds of powerful pumps, an entire sea could be sucked up and redistributed in less than twenty-four hours (see Figure 7).

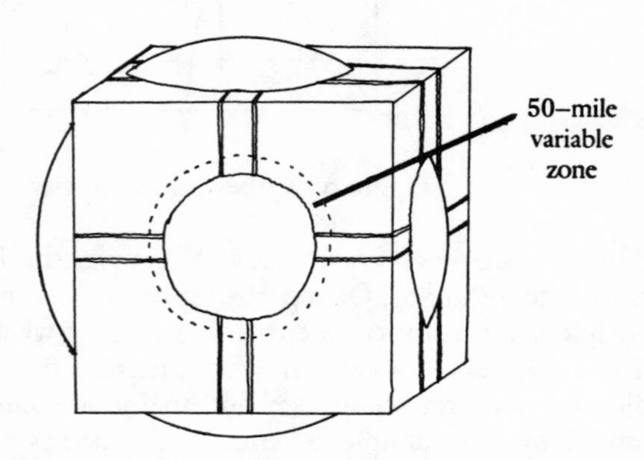

Figure 7 Ocean Relocation System

The awesome power required to pump all this liquid about so quickly derived from the seas themselves. Because all six oceans were perfectly circular, they were subject for the first time to the Coriolis effect—the rotation of bodies of liquid in response to the Earth's rotation, an effect most commonly observed in toilets, which flush in whirlpools, clockwise or counterclockwise, depending upon where on either Earth or Cubeworld they are located. So it now was with the seas. Surf swept shores angularly, as if the oceans were being churned by giant blades at the center. Handily, by intercepting these circular currents at the shore with anything from homemade water wheels to sophisticated turbines, made mobile to adjust to an ocean's changing diameter, we were able to tap all the free, absolutely pollutionless power we needed. In addition, clean, free power could also be teased from the wind, which was constantly sweeping radially from edge toward center, whooshing across the oceans and converging turbulently at their middles.

Getting about on Cubeworld proved to be another pleasant surprise. Initially, as we spread out to reconquer our planet, almost all travel was by vessel. Happily, getting across an ocean required no power whatsoever; you simply drifted around with the Coriolis current until you arrived where you wanted to be and then stepped off as if from a carousel. (One had to avoid the center, of course, which resembled the inside of a Cuisinart!) As airports were rebuilt, air transport again came into its own, thereby facilitating interfacial travel. Following this, we rebuilt our road and railway systems, arranging them in smartly banked spirals to take advantage of Cubeworld's ubiquitous steepness. A bicyclist, for example, could coast for thousands of miles along the Coriolis Cruiseway, our largest vehicular artery, without a single turn of the pedals, as if negotiating the inside of a shallow bowl from rim to bottom. Those hardier souls wishing to make their way from one side to another as pedestrians could do so by following the route of the buried water pipes—climbing up to the edge of the world and savoring the magnificent vista that greeted one upon looking over or, to the same extent, by looking behind oneself. Although the air was thinner in these areas, congenially, it was impossible to fall completely "off" the edge of this world, although you could certainly roll down a fairly steep pitch (forty-five degrees at the outset and gradually diminishing), an opportunity soon seized upon by skateboarders and other daredevils. Most innovative of all, however, was the intercubal express transit network which made dual use of the giant underground water pipes primarily intended to transfer ocean water from one face of Cubeworld to another to counter the Salton Cycle. Passengers could ride in comfortable, pressurized capsules through these aqueducts, powered without cost by the natural ocean currents. Mail, produce, and merchandise were all eventually floated through this network, as well.

These same great pipes, in conjunction with enormous desalinization apparatus, also served to distribute water about the planet for drinking and irrigation purposes. This was necessary because Cubeworld's centrally pulling gravity caused any water that evaporated naturally from the seas' surfaces to concentrate at the centers rather than disperse homogeneously throughout the world. Of course, the plus side of our oceans' miserliness with their H_2O was that the weather on Cubeworld was a picnicker's dream. It was almost always sunny and pleasant, if somewhat windy, which resulted over time in an architectural style emphasizing vertical windbreaks. As rare as natural rain in summer was natural snow in winter, although artificial snow was frequently generated for recreational purposes. When it fell, it did so at an angle to the ground; except when there was a stiff wind, and then it fell "straight down." It wasn't long before some clever Cubeworldian solved the problem of multiangular precipitation by inventing the "All-Latitude Adjustable Umbrella."

Communication—my department—proved to be a cinch, in spite of the fact that almost all ground-wired communication links had been destroyed during transformation and all communications satellites, their orbits outraged by our spiffy new shape, were out of whack and on their way to fiery re-entry. The reason was that we no longer had a big planetary paunch—our old horizon—getting in the way of straight-line communications. As a consequence, we were able to interconnect the entire planet by building reception-transmission towers atop just three of Cubeworld's eight apexes. Because these apexes extended well beyond our atmosphere, the towers built upon them were completely free from meteorological buffeting and hence no longer required massive supports, while a further advantage of being above the atmosphere was that the towers could exchange data at optical wavelengths directly through "outer space" without the need for long glass fibers, thus boosting our information-processing capacity far beyond even the human race's hunger for chatter (see Figure 8).

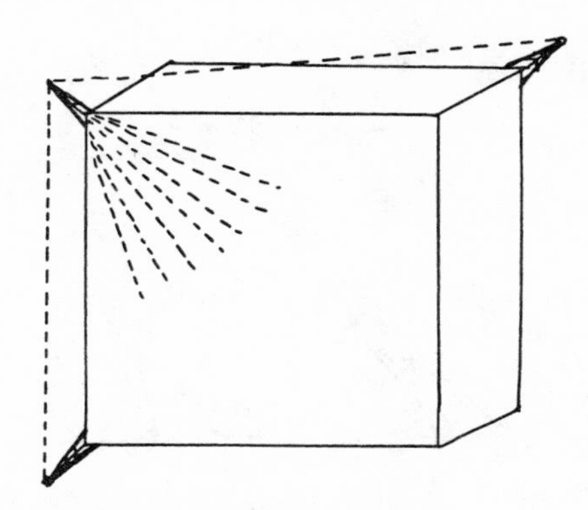

Figure 8 Communications

Of Cubeworld's six available sides, one, called Agriland, was set aside in its entirety as the planet's breadbasket. Thanks to our new capability, it was aimed toward the sun so as to receive twenty hours of light a day without seriously shortchanging the other sides. Soon endless fields of wheat and corn reached, at an angle, to the sky. Rice paddies flourished. Erosion, while ubiquitous in our inclined environment, was turned to advantage, as fresh soil was constantly hauled to the outlying districts and allowed to migrate back to the sea, sharing its nutrients with crops along the way—the old farming technique of crop rotation having given way to the new one of soil rotation. There were thousands of square miles of grazing land, livestock galore, chicken farms, eggeries, and orange groves—whose trees all slanted, as if cringing, and whose fruit, were it to have fallen on some modern Newton's head, would have done so at an angle as well (see Figure 9). It was all almost too good to be believed. Agriland was blessed with unlimited energy, water, and sunlight. As a result of this easy abundance, Rufus Cortez proclaimed food on Cubeworld free to all.

Governmentally, Cubeworld was organized as a pentocracy, with each of the five sides (not counting Agriland) having one vote re-

Figure 9 Sir Murray Newton's Orange

garding cubal policy, three votes constituting a working majority. Should one side abstain and the others split two-two, Rufus Cortez could cast the deciding ballot; in regard to any astronomical emergency, however, he remained our autocratic Chief of Operations. The basic unit of monetary exchange was the cubit, naturally, and a universal flat sales tax provided sufficient funds to meet the common need. The major outlay was for machinery and infrastructure, since power and food were cheap and abundant.

Though we were Pentocrats all, in recognition that it took all kinds to make a world, whatever its shape, Cubeworld's five "residential" sides were further divided according to three levels of freedom: lots of it, almost none of it, and somewhere in between (see Figure 10). The side called Freeland was, by design, the most uninhibited and was settled by those of us to whom a healthy dose of chaos is like a breath of air. It hosted a disproportionate number of artists, musicians, inventors, and all-night dance palaces, as well as a spectacular resort called Artificial Hawaii; some claimed that even the angled trees seemed more laid back here than on the other sides. Here creativity abounded, highways looped and curved gaily, if not illogically, schools gave the least homework of any square on

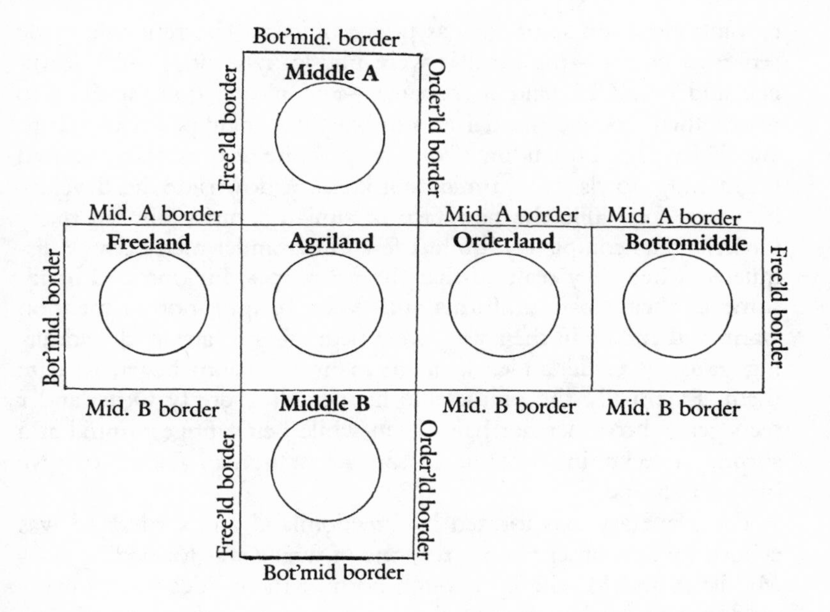

Figure 10 Social Structure

the planet, and gray matter and the eccentric touch were cherished and applauded. Indeed, it was on Freeland that one freewheeling scientist developed a process to make gray matter itself brightly colored, while another researcher discovered a wholly new basic taste, so that the human race, long limited to four, now had five: sweet, sour, salt, bitter, and Walter, which was especially good sprinkled on eggs.

On the opposite side of the planet was Orderland, to whose gridlike burgs gravitated those folks preferring a higher degree of security and predictability in their lives, and where one had to be fingerprinted to buy a slice of pizza, get a haircut, cross a toll bridge, or go to a (censored, of course) movie. The entire side was planned and modular, a facewide grid of greater and lesser squares, right down to the shape of the buildings. So enamored of authority were the repressed people who chose to live on this sixth of our world that their society gave rise to one of mankind's oddest social revolutions: the use of apes as police officers. The rationale made sense, of course—the gorillas were physically strong, quick learners, and could be paid in bananas—and it was quite a sight to watch them receive martial arts instruction, their pendulous arms and hairy legs protruding from crisp white *gis*, as they emitted frightening howls and intimidating kicks. A downside did develop, however, for while the new face of simian security pleased many Orderlanders completely, quite a few Orderlander men became disquieted when they realized that the great apes, hulking and handsome in their smart uniforms and flashy badges, not to mention warm and cuddly in their way, were beginning to attract the admiring gazes of Orderlander women, many of whom began to date them. Eventually the predictable happened: a pretty Orderlander teenager, who'd met her hairy beau while being fingerprinted at a surprise checkpoint, became history's first human female to give birth to an ape.

The planetary axis formed by Freedomland and Orderland was echoed by two other major axes, one of which was formed by sides Middle A and Middle B. Though both of these sides were equally moderate-minded in terms of personal liberties, they nevertheless soon became polarized in that Middle A grew increasingly analog while its opposite side waxed enthusiastically digital. That is to say, those who, for example, liked the hands-on feel of slide rules rather than pocket calculators made their homes and friends mainly on Middle A, while on Middle B, generally speaking, you'd find people whose taste ran to, say, clocks with LED readouts in place of the traditional little hands and big hands ubiquitous on Middle A. This dichotomy extended to hundreds of consumer items, from cameras to bathroom scales, as well as to interpersonal relationships. For example, Middle B men would be apt to measure a woman's beauty digitally, say on a scale of one to ten, while Middle A-ers would instead describe hourglass figures with their hands, waggle their eyebrows, and make appreciative vocal sounds. For this reason, Middle A soon acquired a nickname among some: France.

The planet's third axis was formed by the opposing sides of Agriland and Bottomiddle. That is, in contrast to the rural lushness of Agriland, Bottomiddle grew densely urban, finally knitting into a vast planetary city and thus providing both city lovers and country lovers with a clear-cut choice of environments. Bottomiddle, of course, was so named because, while it was socially as middle-of-the-road as Middles A and B, it was geometrically the rear end of Agriland, and Agriland was the side which—as it faced the light for a longer period than any other side—was universally considered Cubeworld's "top." That is to say, if a giant being striding across the solar system were to pause for a rest, Agriland was the side on which we all instinctively felt it would sit. Certainly not on a corner!

In short, our new world had something for everyone. With unrestricted travel amongst the five "states" cubally guaranteed; with the risk of nuclear self-destruction gone forever; and with everybody's belly full and optimism high, there was no question about it: Cubeworld was full and rich and pleasantly bizarre; six happy valleys (so far as our perception was concerned) sailing cockily through space as our poets romanticized its virtues: its funky new gravity, exemplary weather, bobsled superhighways, and free food. A popular song, which became a hit on all six sides of the new planet, celebrated Cubeworld's ability to tilt: "I like your attitude, I'm on your latitude, you have my gratitude, Cubeworld.'" And that first year, in February, Sarah gave birth to Michael, a lovely boy child who had never known gravity as we had, and now we were four—just like the equilateral horizons of Bottomiddle where we lived.

At last, as predicted, the Salton Mass approached, and we were ready for it. We shunted oceans extravagantly, our planet "passing water," as some jokesters said; and, as all of us know, it worked—though not without some discomfiture, for as Cubeworld's posture in the sky became altered, so did the equilibrium of the entire solar system. It was as if we were inside a planetarium whose projector had gone out of control. Even distant Pluto moved perceptibly into a new orbit, while here at home the lengths of our days stretched and shrank daily, causing chronic jet lag for billions of us. Because the sun no longer rose or set over our curved horizon but rather slid along them as if being sheared by giant scissors, many people became nauseated and had to stay indoors, although some did find the bizarre display entertaining and got high on it. Despite the many virtues of a cubic world, a "Round Is Sound" movement developed whose adherents wanted to restore the planet to a spherical shape now that the danger from the Salton Mass had passed.

As things turned out, we were forced by the power of geometry itself to do just that.

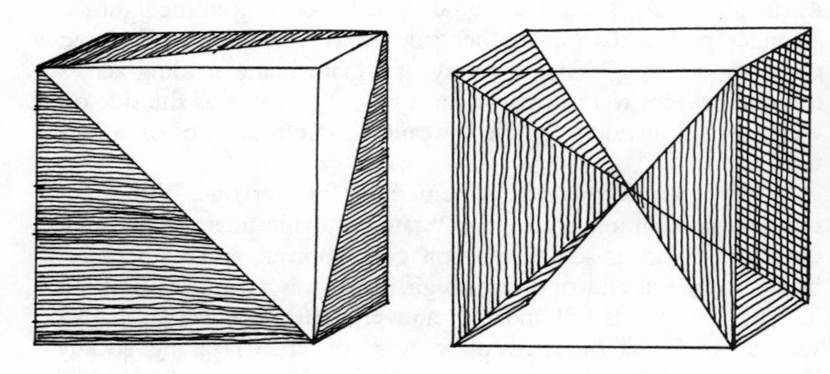

Figure 11 Three-sided Major Pyramids and Four-sided Minor Pyramids

The problem was that a cube, as can be seen in Figure 11, is composed, implicitly, of eight major three-sided pyramids, each with an apex at a corner of the cube and a triangular base inside the cube, plus six minor four-sided pyramids with bases formed by the cube's six sides and apexes converging at the center. And when our scientists first detected seismological stresses in Cubeworld's interior which threatened to blow the entire planet to smithereens, they speculated—once all known physical phenomena had been eliminated as possible causes—that the stresses were caused by these massive inherent pyramids generating and "focusing" at our planet's center enormous amounts of "pyramid power"—the hypothetical energy that supposedly keeps mummies fresh for millennia, improves general health, makes seeds sprout faster, and sharpens razor blades. Unorthodox as this notion was, it was lent credence by the fact that our population was indeed healthier, seeds were indeed bursting forth more vigorously, and yes, the razor industry was in trouble because men inexplicably just weren't buying new blades anymore. Indeed, as Chou Mao Li, our gentlemanly Chief of Security suggested, cubic planets, perhaps many times the size of our own with correspondingly larger ingrown pyramids, might at some time have been used as giant interstellar space mines in some awesome galactic conflict—or might yet be the ultimate weapons of the future.

Whatever the merits of these theories, the test results were beyond question: Cubeworld had a tiger in its tank, and unless we turned it back into a sphere in a hurry, our cheery little polyhedron would soon be vying with the asteroid belt for most interesting collection of rocks in the solar system. We were truly at the mercy of geometry. Cubeworld was not, after all, the shape of things to come. The postulates of Euclid ruled our universe indeed.

And so, for the second time, we gathered our will and resources and set about to accomplish an epic task of earthmoving. Once again we prepared a month of survival supplies for the world's population. Once again we assigned everyone a task, a position, and a destination. Finally, all was ready; and as our beloved cube gurgled inside with the pyramidal heartburn that threatened to destroy it, we said a collective, televised good-bye to our comfy little planet and began the transformation.

It was as before, only in reverse. Machinery moved into action. Calculated explosions rent the air. Wheelbarrows rolled and supervisors barked. Apexes became domes became hills became gently arcing plains. The great globules of oceans were transformed back into rivers, lakes, and irregularly shaped seas. Bottomiddle was sculpted into Australia, Agriland into Europe, and Freeland into North America, while Middles A and B and Orderland were distributed as needed. The Grand Canyon was re-excavated and Mount Everest re-created and recrowned. Once again our now five point five billion inhabitants spent a month in a global steambath.

At last, as before, the job was done. Gradually the steam dissipated. Sarah, myself, and the kids, perched on a knoll beside the soon to be obsolete Planned Planethood headquarters, looked around us gratefully. Everything was back in place. Indeed, and incredibly, *exactly* back in place. The sun was just beginning to appear over the nearly straightedge (or so it now appeared) horizon to the east; indeed, its appearance is what *made* it east. The temperature was exactly sixty-eight degrees Fahrenheit with low humidity—an absolutely gorgeous morning. And, as I looked off to one side, by golly if that interface between sculpted land and sparkling sea wasn't a dead ringer for the coastline of New Jersey!

Soon we would begin refurbishing our lives to readapt to an environment many of us had quite forgotten during the short time we were cubical. Maybe, indeed, we'd learned something from the experience. The Earth was again spherical, and perhaps that was the way things were supposed to be. Who were we mere humans to argue with the majesty of angle and shape? Geometry was the grand master of the universe after all.

I looked around to see our youngest child, who'd been born on Cubeworld and who'd never experienced existence on a round planet before, running in excitement to Sarah. "Mommy, Mommy!" he called in wonder. "This is weird!"

"What, honey?" she asked, with a happy look toward me.

"It's flat, Mommy! The whole world is flat!"

FREDERIK POHL

Schematic Man

I know I'm not really a funny man, but I don't like other people to know it. I do what other people without much sense of humor do: I tell jokes. If we're sitting next to each other at a faculty senate and I want to introduce myself, I probably say: "Bederkind is my name, and computers are my game."

Nobody laughs much. Like all my jokes, it needs to be explained. The joking part is that it was through game theory that I first became interested in computers and the making of mathematical models. Sometimes when I'm explaining it, I say there that the mathematical ones are the only models I've ever had a chance to make. That gets a smile, anyway. I've figured out why: Even if you don't really get much out of the play on words, you can tell it's got something to do with sex, and we all reflexively smile when anybody says anything sexy.

I ought to tell you what a mathematical model is, right? All right. It's simple. It's a kind of picture of something made out of numbers. You use it because it's easier to make numbers move than to make real things move.

Suppose I want to know what the planet Mars is going to do over the next few years. I take everything I know about Mars and I turn it into numbers—a number for its speed in orbit, another number for how much it weighs, another number for how many miles it is in diameter, another number to express how strongly the sun pulls it towards it and all that. Then I tell the computer that's all it needs to know about Mars, and I go on to tell it all the same sorts of numbers about the Earth, about Venus, Jupiter, the sun itself—about all the other chunks of matter floating around in the neighborhood that I think are likely to make any difference to Mars. I then teach the computer some simple rules about how the set of numbers that represent Jupiter say, affect the numbers that represent Mars: the law of inverse squares, some rules of celestial mechanics, a few relativistic corrections—well, actually, there are a lot of things it needs to know. But not more than I can tell it.

When I have done all this—not exactly in English but in a kind of language that it knows how to handle—the computer has a mathematical model of Mars stored inside it. It will then whirl its mathematical Mars through mathematical space for as many orbits as I like. I say to it, "1997 June 18 2400 GMT," and it . . . it . . . well, I guess the word for it is, it *imagines* where Mars will be, relative to my backyard Questar, at midnight Greenwich time on the 18th of June, 1997, and tells me which way to point.

It isn't real Mars that it plays with. It's a mathematical model, you see. But for the purposes of knowing where to point my little telescope, it does everything that "real Mars" would do for me, only much faster. I don't have to wait for 1997; I can find out in five minutes.

It isn't only planets that can carry on a mathematical metalife in the memory banks of a computer. Take my friend Schmuel. He has a joke, too, and his joke is that he makes twenty babies a day in his computer. What he means by that is that, after six years of trying, he finally succeeded in writing down the numbers that describe the development of a human baby in its mother's uterus, all the way from conception to birth. The point of that is that then it was comparatively easy to write down the numbers for a lot of things that happen to babies before they're born. Momma has high blood pressure. Momma smokes three packs a day. Momma catches scarlet fever or a kick in the belly. Momma keeps making it with Poppa every night until they wheel her into the delivery room. And so on. And the point of *that* is that this way, Schmuel can see some of the things that go wrong and make some babies get born retarded, or blind, or with retrolental fibroplasia or an inability to drink cow's milk. It's easier than sacrificing a lot of pregnant women and cutting them open to see.

OK, you don't want to hear any more about mathematical models, because what kicks are there in mathematical models for you?

I'm glad you asked. Consider a for instance. For instance, suppose last night you were watching the "Late, Late" and you saw Carole Lombard, or maybe Marilyn Monroe with that dinky little skirt blowing up over those pretty thighs. I assume you know that these ladies are dead. I also assume that your glands responded to those cathode-tube flickers as though they were alive. And so you do get some kicks from mathematical models, because each of those great girls, in each of their poses and smiles, was nothing but a number of some thousands of digits, expressed as a spot of light on a phosphor tube. With some added numbers to express the frequency patterns of their voices. Nothing else.

And the point of *that* (how often I use that phrase!) is that a mathematical model not only represents the real thing but sometimes it's as good as the real thing. No, honestly. I mean, do you really believe that if it had been Marilyn or Carole in the flesh you were looking at, across a row of floodlights, say, that you could have taken away any more of them than you gleaned from the shower of electrons that made the phosphors display their pictures?

I did watch Marilyn on the "Late, Late" one night. And I thought those thoughts; and so I spent the next week preparing an application to a foundation for money; and when the grant came through, I took a sabbatical and began turning myself into a mathematical model. It isn't really that hard. Kookie, yes. But not hard.

I don't want to explain what programs like Fortran and Simscript and Sir are, so I will only say what we all say: They are languages by which people can communicate with machines. Sort of. I had to learn to speak Fortran well enough to tell the machine all about myself. It took five graduate students and ten months to write the program that made that possible, but that's not much. It took more than that to teach a computer to shoot pool. After that, it was just a matter of storing myself in the machine.

That's the part that Schmuel told me was kookie. Like everybody with enough seniority in my department, I have a remote-access computer console in my—well, I called it my "playroom." I did have a party there, once, right after I bought the house, when I still thought I was going to get married. Schmuel caught me one night walking in the door and down the stairs and finding me methodically typing out my medical history from the ages of four to fourteen. "Jerk," he said, "what makes you think you deserve to be embalmed in a 7094?"

I said, "Make some coffee and leave me alone till I finish. Listen. Can I use your program on the sequelae of mumps?"

"Paranoid psychosis," he said. "It comes on about the age of forty-two." But he coded the console for me and thus gave me access to his programs. I finished and said:

"Thanks for the program, but you make rotten coffee."

"You make rotten jokes. You really think it's going to be *you* in that program. Admit!"

By then, I had most of the basic physiological and environmental stuff on the tapes and I was feeling good. "What's me?" I asked. "If it talks like me, and thinks like me, and remembers what I remember, and does what I would do—who is it? President Eisenhower?"

"Eisenhower was years ago, jerk," he said.

"Turing's question, Schmuel," I said. "If I'm in one room with a teletype. And the computer's in another room with a teletype, programmed to model me. And you're in a third room, connected to both teletypes, and you have a conversation with both of us, and you can't tell which is me and which is the machine—then how do you describe the difference? *Is* there a difference?"

He said, "The difference, Josiah, is I can touch you. And smell you. If I was crazy enough I could kiss you. You. Not the model."

"You could," I said, "if you were a model, too, and were in the machine with me." And I joked with him (Look! It solves the population problem, put everybody in the machine. And, suppose I get cancer. Flesh-me dies. Mathematical-model-me just rewrites its program), but he was really worried. He really did think I was going crazy, but I perceived that his reasons were not because of the nature of the problem but because of what he fancied was my own attitude towards it, and I made up my mind to be careful of what I said to Schmuel.

So I went on playing Turing's game, trying to make the computer's responses indistinguishable from my own. I instructed it in what a toothache felt like and what I remembered of sex. I taught it memory links between people and phone numbers, and all the state capitals I had won a prize for knowing when I was ten. I trained it to spell "rhythm" wrong, as I had always misspelled it, and to say "place" instead of "put" in conversation, as I have always done because of the slight speech impediment that carried over from my adolescence. I played that game; and by God, I won it.

But I don't know for sure what I lost in exchange.

I know I lost something.

I began by losing parts of my memory. When my cousin Alvin from Cleveland phoned me on my birthday, I couldn't remember who he was for a minute. (The week before, I had told the computer all about my summers with Alvin's family, including the afternoon when we both lost our virginity to the same girl, under the bridge by my uncle's farm.) I had to write down Schmuel's phone number, and my secretary's, and carry them around in my pocket.

As the work progressed, I lost more. I looked up at the sky one night and saw three bright stars in a line overhead. It scared me, because I didn't know what they were until I got home and took out my sky charts. Yet Orion was my first and easiest constellation. And when I looked at the telescope I had made, I could not remember how I had figured the mirror.

Schmuel kept warning me about overwork. I really was working a lot, fifteen hours a day and more. But it didn't feel like overwork. It felt as though I were losing pieces of myself. I was not merely teaching the computer to be me but putting pieces of me into the computer. I hated that, and it shook me enough to make me take the whole of Christmas week off. I went to Miami.

But when I got back to work, I couldn't remember how to touch-type on the console any more and was reduced to pecking out information for the computer a letter at a time. I felt as though I were moving from one place to another in installments, and not enough of me had arrived yet to be a quorum, but what was still waiting to go had important parts missing. And yet I continued to pour myself into the magnetic memory cores: the lie I told my draft board in 1946, the limerick I made up about my first wife after the divorce, what Margaret wrote when she told me she wouldn't marry me.

There was plenty of room in the storage banks for all of it. The computer could hold all my brain had held, especially with the program my five graduate students and I had written. I had been worried about that, at first.

But in the event I did not run out of room. What I ran out of was myself. I remember feeling sort of opaque and stunned and empty; and that is all I remember until now.

Whenever "now" is.

I had another friend once, and he cracked up while working on telemetry studies for one of the Mariner programs. I remember going to see him in the hospital, and him telling me, in his slow, unworried, coked-up voice, what they had done for him. Or to him. Electroshock. Hydrotherapy.

What worries me is that that is at least a reasonable working hypothesis to describe what is happening to me now.

I remember, or think I remember, a sharp electric jolt. I feel, or think I feel, a chilling flow around me.

What does it mean? I wish I were sure. I'm willing to concede that it might mean that overwork did me in and now I, too, am at Restful Retreat, being studied by the psychiatrists and changed by the nurses' aides. Willing to concede it? Dear God, I *pray* for it. I pray that electricity was just shock therapy and not something else. I pray that the flow I feel is water sluicing around my sodden sheets and not a flux of electrons in transistor modules. I don't fear the thought of being insane; I fear the alternative.

I do not *believe* the alternative. But I fear it all the same. I can't believe that all that's left of me—my id, my ucs, my *me*—is nothing but a mathematical model stored inside the banks of the 7094. But if I am! If I am, dear God, what will happen when—and how can I wait until—somebody turns me on?

GREGORY BENFORD

Time's Rub

1

At Earth's winter ebb, two crabbed figures slouched across a dry, cracked plain.

Running before a victor who was himself slow-dying, the dead stench of certain destiny cloyed to them. They knew it. Yet kept on, grinding over plum-colored shales.

They shambled into a pitwallow for shelter, groaning, carapaces grimed and discolored. The smaller of them, Xen, turned toward the minimal speck of burnt-yellow sun, but gained little aid through its battered external panels. It grasped Faz's extended pincer—useless now, mauled in battle—and murmured of fatigue.

"We can't go on."

Faz, grimly: "We must."

Xen was a functionary, an analytical sort. It had chanced to flee the battle down the same gully as Faz, the massive, lumbering leader. Xen yearned to see again its mate, Pymr, but knew this for the forlorn dream it was.

They crouched down. Their enemies rumbled in nearby ruined hills. A brown murk rose from those distant movements. The sun's pale eye stretched long shadows across the plain, inky hiding places for the encroaching others.

Thus when the shimmering curtains of ivory luminescence began to fog the hollow, Xen thought the end was here—that energy drain blurred its brain, and now brought swift, cutting death.

Fresh in from the darkling plain? the voice said. Not acoustically—this was a Vac Zone, airless for millennia.

"What? Who's that?" Faz answered.

Your ignorant armies clashed last night?

"Yes," Xen acknowledged ruefully, "and were defeated. Both sides lost."

Often the case.

"Are the Laggenmorphs far behind us?" Faz asked, faint tracers of hope skating crimson in its spiky voice.

No. They approach. They have tracked your confused alarms of struggle and flight.

"We had hoped to steal silent."

Your rear guard made a melancholy, long, withdrawing roar.

Xen: "They escaped?"

Into the next world, yes.

"Oh."

"Who *is* that?" Faz insisted, clattering its treads.

A wraith. Glittering skeins danced around them. A patchy acrid tang laced the curling vacuum. **In this place having neither brass, nor earth, nor boundless sea.**

"Come out!" Faz called at 3 gigahertz. "We can't see you."

Need you?

"Are you Laggenmorphs?" Panic shaded Faz's carrier wave a bright, fervid orange. "We'll fight, I warn you!"

"Quiet," Xen said, suspecting.

The descending dazzle thickened, struck a bass note. **Laggenmorphs? I do not even know your terms.**

"Your name, then," Xen said.

Sam.

"What's that? That's no name!" Faz declared, its voice a shifting brew of fear and anger.

Sam it was and Sam it is. Not marble, nor the gilded monuments of princes, shall outlive it.

Xen murmured at 100 kilohertz, "Traditional archaic name. I dimly remember something of the sort. I doubt it's a trap."

The words not yet free of its antenna, Xen ducked—for a relativistic beam passed not a kilometer away, snapping with random rage. It forked to a ruined scree of limestone and erupted into a self-satisfied yellow geyser. Stones pelted the two hunkering forms, clanging.

A mere stochastic volley. Your sort do expend energies wildly. That is what first attracted me.

Surly, Faz snapped. "You'll get no surge from us."

I did not come to sup. I came to proffer.

A saffron umbra surrounded the still-gathering whorls of crackling, clotted iridescence.

"Where're you hiding?" Faz demanded. It brandished blades, snouts, cutters, spikes, double-bore nostrils that could spit lurid beams.

In the cupped air.

"There *is* no air," Xen said. "This channel is open to the planetary currents."

Xen gestured upward with half-shattered claw. There, standing in space, the playing tides of blue-white, gauzy light showed that they were at the base of a great translucent cylinder. Its geometric perfection held back the moist air of Earth, now an ocean tamed by skewered forces. On the horizon, at the glimmering boundary, purpling clouds nudged futilely at their constraint like hungry cattle. This cylinder led the eye up to a vastness, the stars a stilled snowfall. Here the thin but persistent wind from the sun could have free run, gliding along the orange-slice sections of the Earth's dipolar magnetic fields. The winds crashed down, sputtering, delivering kilovolt glories where the cylinder cut them. Crackling yellow sparks grew there, a forest with all trunks ablaze and branches of lightning, beckoning far aloft like a brilliantly lit casino in a gray dark desert.

How well I know. I stem from fossiled days.

"Then why—"

This is my destiny and my sentence.

"To live here?" Faz was beginning to suspect as well.

For a wink or two of eternity.

"Can you . . ." Faz poked the sky with a horned, fused launcher. ". . . reach up there? Get us a jec?"

I do not know the term.

Xen said, "An injection. A megavolt, say, at a hundred kiloamps. A mere microsecond would boost me again. I could get my crawlers working."

I would have to extend my field lines.

"So it *is* true," Xen said triumphantly. "There still dwell Ims on the Earth. And you're one."

Again, the term—

"An Immortal. You have the fieldcraft."

Yes.

Xen knew of this, but had thought it mere legend. All material things were mortal. Cells were subject to intruding impurities, cancerous insults, a thousand coarse alleyways of accident. Machines, too, knew rust and wear, could suffer the ruthless scrubbing of their memories by a random bolt of electromagnetic violence. Hybrids, such as Xen and Faz, shared both half-worlds of erosion.

But there was a Principle which evaded time's rub. Order could be imposed on electrical currents—much as words rode on radio waves—and then the currents could curve into self-involved equilibria. It spun just so, the mouth of a given stream eating its own tail, then a spinning ring generated its own magnetic fields. Such work was simple. Little children made these loops, juggled them into humming fireworks.

Only genius could knit these current whorls into a fully contorted globe. The fundamental physics sprang from ancient Man's bottling of thermonuclear fusion in magnetic strands. That was a simple craft, using brute magnets and artful metallic vessels. Far harder, to apply such learning to wisps of plasma alone.

The Principle stated that if, from the calm center of such a weave, the magnetic field always increased, in all directions, then it was stable to all manner of magnetohydrodynamic pinches and shoves.

The Principle was clear, but stitching the loops—history had swallowed that secret. A few had made the leap, been translated into surges of magnetic field. They dwelled in the Vac Zones, where the rude bump of air molecules could not stir their calm currents. Such were the Ims.

"You . . . live forever?" Xen asked wonderingly.

Aye, a holy spinning toroid—when I rest. Otherwise, distorted, as you see me now. Phantom shoots of burnt yellow. **What once was Man, is now aurora—where winds don't sing, the sun's a tarnished nickel, the sky's a blank rebuke.**

Abruptly, a dun-colored javelin shot from nearby ruined hills, vectoring on them.

"Laggenmorphs!" Faz sent. "I have no defense."

Halfway to them, the lance burst into scarlet plumes. The flames guttered out.

A cacophony of eruptions spat from their left. Gray forms leapt forward, sending scarlet beams and bursts. Sharp metal cut the smoking stones.

"Pymr, sleek and soft, I loved you," Xen murmured, thinking this was the end.

But from the space around the Laggenmorphs condensed a chalky stuff—smothering, consuming. The forms fell dead.

I saved you.

Xen bowed, not knowing how to thank a wisp. But the blur of nearing oblivion weighed like stone.

"Help us!" Faz's despair lanced like pain through the dead vacuum. "We need energy."

You would have me tick over the tilt of Earth, run through solstice, bring ringing summer in an hour?

Xen caught in the phosphorescent stipple a green underlay of irony.

"No, no!" Faz spurted. "Just a jec. We'll go on then."

I can make you go on forever.

The flatness of it; accompanied by phantom shoots of scorched orange, gave Xen pause. "You mean . . . the fieldcraft? Even I know such lore is not lightly passed on. Too many Ims, and the Earth's magnetic zones will be congested."

I grow bored, encased in this glassy electromagnetic shaft. I have not conferred the fieldcraft in a long while. Seeing you come crawling from your mad white chaos, I desired company. I propose a Game.

"Game?" Faz was instantly suspicious. "Just a jec, Im, that's all we want."

You may have that as well.

"What're you spilling about?" Faz asked.

Xen said warily, "It's offering the secret."

"What?" Faz laughed dryly, a flat cynical burst that rattled down the frequencies.

Faz churned an extruded leg against the grainy soil, wasting energy in its own consuming bitterness. It had sought fame, dominion, a sliver of history. Its divisions had been chewed and spat out again by the Laggenmorphs, its feints ignored, bold strokes adroitly turned aside. Now it had to fly vanquished beside the lesser Xen, dignity gathered like tattered dress about its fleeing ankles.

"Ims never share *that*. A dollop, a jec, sure—but not the turns of fieldcraft." To show it would not be fooled, Faz spat chalky ejecta at a nearby streamer of zinc-laden light.

I offer you my Game.

The sour despair in Faz spoke first. "Even if we believe that, how do we know you don't cheat?"

No answer. But from the high hard vault there came descending a huge ribbon of ruby light—snaking, flexing, writing in strange tongues on the emptiness as it approached, fleeting messages of times gone—auguries of innocence lost, missions forgot, dim songs of the wide world and all its fading sweets. The ruby snake split, rumbled, turned eggshell blue, split and spread and forked down, blooming into a hemisphere around them. It struck and ripped the rock, spitting fragments over their swiveling heads, booming. Then prickly silence.

"I see," Xen said.

Thunder impresses, but it's lightning does the work.

"Why should the Im cheat, when it could short us to ground, fry us to slag?" Xen sent to Faz on tightband.

"Why anything?" Faz answered, but there was nodding in the tone.

2

The Im twisted the local fields and caused to appear, hovering in fried light, two cubes—one red, one blue.

You may choose to open either the Blue cube alone, or both.

Though brightened by a borrowed kiloamp jolt from Xen, Faz had expended many joules in irritation and now flagged. "What's . . . in . . . them?"

Their contents are determined by what I have already predicted. I have already placed your rewards inside. You can choose Red and Blue both, if you want. In that case, following my prediction, I have placed in the Red cube the bottled-up injection you wanted.

Faz unfurled a metallic tentacle for the Red cube.

Wait. If you will open both boxes, then I have placed in the Blue nothing—nothing at all.

Faz said, "Then I get the jec in the Red cube, and when I open the Blue—nothing."

Correct.

Xen asked, "What if Faz *doesn't* open both cubes?"

The only other option is to open the Blue alone.

"And I get nothing?" Faz asked.

No. In that case, I have placed the, ah, "jec" in the Red cube. But in the Blue I have put the key to my own fieldcraft—the designs for immortality.

"I don't get it. I open Red, I get my jec—right?" Faz said, sudden interest giving it a spike of scarlet brilliance at 3 gigahertz. "Then I open Blue, I get immortality. That's what I want."

True. But in that case, I have predicted that you will pick both cubes. Therefore, I have left the Blue cube empty.

Faz clattered its treads. "I get immortality if I choose the Blue cube *alone*? But you have to have *predicted* that. Otherwise I get nothing."

Yes.

Xen added. "*If* you have predicted things perfectly."

But I always do.

"Always?"

Nearly always. I am immortal, ageless—but not God. Not . . . yet.

"What if I pick Blue and you're wrong?" Faz asked. "Then I get nothing."

True. But highly improbable.

Xen saw it. "All this is done *now*? You've already made your prediction? Placed the jec or the secret—or both—in the cubes?"

Yes. I made my predictions before I even offered the Game.

Faz asked, "What'd you predict?"

Merry pink laughter chimed across the slumbering mega-Hertz. **I will not say. Except that I predicted correctly that you both would play, and that you particularly would ask that question. Witness.**

A sucking jolt lifted Faz from the stones and deposited it nearby. Etched in the rock beneath where Faz had crouched was *What did you predict?* in a rounded, careful hand.

"It had to have been done during the overhead display, before the game began," Xen said wonderingly.

"The Im *can* predict," Faz said respectfully.

Xen said, "Then the smart move is to open both cubes."

Why?

"Because you've already made your choice. If you predicted that Faz would choose both, and he opens only the Blue, then he gets nothing."

True, and as I said before, very improbable.

"So," Xen went on, thinking quickly under its pocked sheen of titanium, "if you predicted that Faz would choose *only* the Blue, then Faz might as well open both. Faz will get both the jec and the secret."

Faz said, "Right. And that jec will be useful in getting away from here."

Except that there is every possibility that I already predicted his choice of both cubes. In that case I have left only the jec in the Red Cube, and nothing in Blue.

"But you've already chosen!" Faz blurted. "There isn't any probable-this or possible-that at all."

True.

Xen said, "The only uncertainty is, how good a predictor are you."

Quite.

Faz slowed, flexing a crane arm in agonized frustration. "I . . . dunno . . . I got . . . to think . . ."

There's world enough, and time.

"Let me draw a diagram," said Xen, who had always favored the orderly over the dramatic. This was what condemned it to a minor role in roiling battle, but perhaps that was a blessing. It drew upon the gritty soil some boxes: "There," Xen wheezed. "This is the payoff matrix."

As solemn and formal as Job's argument with God.

Enraptured with his own creation, Xen said, "Clearly, taking only the Blue cube is the best choice. The chances that the Im are wrong are very small. So you have a great chance of gaining immortality."

"That's crazy," Faz mumbled. "If I take both cubes, I at *least* get a jec, even if the Im *knew* I'd choose that way. And with a jec, I can make a run for it from the Laggenmorphs."

"Yes. Yet it rests on faith," Xen said. "Faith that the Im's predicting is near-perfect."

"Ha!" Faz snorted. "Nothing's perfect."

A black thing scorched over the rim of the pitwallow and ex-

		THE IM: Predicts you will take only what's in Blue	THE IM: Predicts you will take what's in both
YOU	Take only what is in Blue	immortality	nothing
	Take what is in both Red and Blue	immortality and jec	jec

ploded into fragments. Each bit dove for Xen and Faz, like shrieking, elongated eagles baring teeth.

And each struck something invisible but solid. Each smacked like an insect striking the windshield of a speeding car. And was gone.

"They're all around us!" Faz cried.

"Even with a jec, we might not make it out," Xen said.

True. But translated into currents, like me, with a subtle knowledge of conductivities and diffusion rates, you can live forever.

"Translated . . ." Xen mused.

Free of entropy's swamp.

"Look," Faz said, "I may be tired, drained, but I know logic. You've already *made* your choice, Im—the cubes are filled with whatever you put in. What I choose to do now can't change that. So I'll take *both* cubes."

Very well.

Faz sprang to the cubes. They burst open with a popping ivory radiance. From the red came a blinding bolt of a jec. It surrounded Faz's antennae and cascaded into the creature.

Drifting lightly from the blue cube came a tight-wound thing, a shifting ball of neon-lit string. Luminous, writhing rainbow worms. They described the complex web of magnetic field geometries that were immortality's craft. Faz seized it.

You won both. I predicted you would take only the blue. I was wrong.

"Ha!" Faz whirled with renewed energy.

Take the model of the fieldcraft. From it you can deduce the methods.

"Come on, Xen!" Faz cried with sudden ferocity. It surged over the lip of the pitwallow, firing at the distant, moving shapes of the Laggenmorphs, full once more of spit and dash. Leaving Xen.

"With that jec, Faz will make it."

I predict so, yes. You could follow Faz. Under cover of its armory, you would find escape—that way.

The shimmer vectored quick a green arrow to westward, where clouds billowed white. There the elements still governed and mortality walked.

"My path lies homeward, to the south."

Bound for Pymr.

"She is the one true rest I have."

You could rest forever.

"Like you? Or Faz, when it masters the . . . translation?"

Yes. Then I will have company here.

"Aha! That is your motivation."

In part.

"What else, then?"

There are rules for immortals. Ones you cannot understand . . . yet.

"If you can predict so well, with godlike power, then I should choose only the Blue cube."

True. Or as true as true gets.

"But if you predict so well, my 'choice' is mere illusion. It was foreordained."

That old saw? I can see you are . . . determined . . . to have free will.

"Or free won't."

Your turn.

"There are issues here . . ." Xen transmitted only ruby ruminations, murmuring like surf on a distant shore.

Distant boomings from Faz's retreat. The Red and Blue cubes spun, sparkling, surfaces rippled by ion-acoustic modes. The game had been reset by the Im, whose curtains of gauzy green shimmered in anticipation.

There must be a Game, you see.

"Otherwise there is no free will?"

That is indeed one of our rules. Observant, you are. I believe I will enjoy the company of you, Xen, more than that of Faz.

"To be . . . an immortal . . ."

A crystalline paradise, better than blind Milton's scribbled vision.

A cluster of dirty-brown explosions ripped the sky, rocked the land.

I cannot expend my voltages much longer. Would that we had wit enough, and time, to continue this parrying.

"All right." Xen raised itself up and clawed away the phosphorescent layers of both cubes.

The Red held a shimmering jec.

The Blue held nothing.

Xen said slowly, "So you predicted correctly."

Yes. Sadly, I knew you too well.

Xen radiated a strange sensation of joy, unlaced by regret. It surged to the lip of the crumbling pitwallow.

"Ah . . ." Xen sent a lofting note. "I am like a book, old Im. No doubt I would suffer in translation."

A last glance backward at the wraith of glow and darkness, a gesture of salute, then: "On! To sound and fury!" and it was gone forever.

3

In the stretched silent years there was time for introspection. Faz learned the lacy straits of Earth's magnetic oceans, its tides and times. It sailed the magnetosphere and spoke to stars.

The deep-etched memories of that encounter persisted. It never saw Xen again, though word did come vibrating through the field lines of Xen's escape, of zestful adventures out in the raw territory of air and Man. There was even a report that Xen had itself and Pymr decanted into full Manform, to taste the pangs of cell and membrane. Clearly, Xen had lived fully after that solstice day. Fresh verve had driven that blithe new spirit.

Faz was now grown full, could scarcely be distinguished from the Im who gave the fieldcraft. Solemn and wise, its induction, conductivity, and ruby glinting dielectrics a glory to be admired, it hung vast and cold in the sky. Faz spoke seldom and thought much.

Yet the game still occupied Faz. It understood with the embedded viewpoint of an immortal now, saw that each side in the game paid a price. The Im could convey the fieldcraft to only a few, and had nearly exhausted itself; those moments cost millennia.

The sacrifice of Faz was less clear.

Faz felt itself the same as before. Its memories were stored in Alfven waves—stirrings of the field lines, standing waves between Earth's magnetic poles. They would be safe until Earth itself wound down, and the dynamo at the nickel-iron core ceased to replenish the fields. Perhaps, by that time, there would be other field lines threading Earth's, and the Ims could spread outward, blending into the galactic fields.

There were signs that such an end had come to other worlds. The cosmic rays which sleeted down perpetually were random, isotropic, which meant they had to be scattered from magnetic waves between the stars. If such waves were ordered, wise—it meant a vast community of even greater Ims.

But this far future did not concern Faz. For it, the past still sang, gritty and real.

Faz asked the Im about that time, during one of their chance auroral meetings, beside a cascading crimson churn.

The way we would put it in my day, the Im named Sam said, **would be that the software never knows what the original hardware was.**

And that was it, Faz saw. During the translation, the original husk of Faz had been exactly memorized. This meant determining the exact locations of each atom, every darting electron. By the quantum laws, to locate perfectly implied that the measurement imparted an unknown, but high, momentum to each speck. So to define a thing precisely then destroyed it.

Yet there was no external way to prove this. Before and after translation there was an exact Faz.

The copy did not know it was embedded in different . . . hardware . . . than the original.

So immortality was a concept with legitimacy purely seen from the outside. From the inside . . .

Somewhere, a Faz had died that this Faz might live.

. . . And how did any sentience know it was not a copy of some long-gone original?

One day, near the sheath that held back the atmosphere, Faz saw a man waving. It stood in green and vibrant wealth of life, clothed at the waist, bronzed. Faz attached a plasma transducer at the boundary and heard the figure say, "You're Faz, right?"

Yes, in a way. And you. . . ?

"Wondered how you liked it."

Xen? Is that you?

"In a way."

You knew.

"Yes. So I went in the opposite direction—into this form."

You'll die soon.

"You've died already."

Still, in your last moments, you'll wish for this.

"No, it's not how long something lasts, it's what that something means." With that the human turned, waved gaily, and trotted into a nearby forest.

This encounter bothered Faz.

In its studies and learned colloquy, Faz saw and felt the tales of Men. They seemed curiously convoluted, revolving about Self. What mattered most to those who loved tales was how they concluded. Yet all Men knew how each ended. Their little dreams were rounded with a sleep.

So the point of a tale was not how it ended, but *what it meant*. The great inspiring epic rage of Man was to find that lesson, buried in a grave.

As each year waned, Faz reflected, and knew that Xen had seen this point. Immortality seen from without, by those who could not know the inner Self—Xen did not want that. So it misled the Im, and got the mere jec that it wanted.

Xen chose life—not to be a monument of unaging intellect, gathered into the artifice of eternity.

In the brittle night Faz wondered if it had chosen well itself. And knew. *Nothing* could be sure it was itself the original. So the only intelligent course lay in enjoying whatever life a being felt—living like a mortal, in the moment. Faz had spent so long, only to reach that same conclusion which was forced on Man from the beginning.

Faz emitted a sprinkling of electromagnetic tones, spattering rueful red the field lines.

And stirred itself to think again, each time the dim sun waned at the solstice. To remember and, still living, to rejoice.

RUDY RUCKER

Message Found in a Copy of *Flatland*

The story, which appears below, is purported to be Robert Ackley's first-person account of his strange disappearance. I am not quite sure if the account is really true . . . I rather hope, for Ackley's sake, that it is not.

I obtained the typescript of this story in a round-about way. My friend, Gregory Gibson, was in London last year, looking for rare books. A dealer in Cheapside showed Gibson a copy of an early edition of Edwin Abbott's 1884 fiction, Flatland. *The copy Gibson saw was remarkable for the fact that someone had handwritten a whole story in the margins of the book's pages. The dealer told Gibson that the volume was brought in by a cook's helper, who had found the book in the basement of a Pakistani restaurant where he once worked.*

Gibson could not afford the book's very steep purchase price, but he did obtain the dealer's permission to copy out the story written in the volume's margins. Here, without further ado, it is: the singular adventure of Robert Ackley.

All my attempts to get back through the tunnel have proven fruitless. It will be necessary for me to move on and seek another way out. Before departing, however, I will write out an account of my adventures thus far.

Until last year I had always believed Edwin Abbott's *Flatland* to be a work of fiction. Now I know better. Flatland is real. I can look up and see it as I write.

For those of my readers familiar with the book in whose margins I write, this will be startling news; for *Flatland* tells the adventures of A Square, a two-dimensional being living in a two-dimensional world. How, you may ask, could such a filmy world really exist? How could there be intelligent creatures with length and width, yet without thickness? If Flatland is real, then why am I the only living man who has touched it? Patience, dear readers. All this, and much more, will be revealed.

The scientific justification for *Flatland* is that it helps us better to understand the fourth dimension. "The fourth dimension" is a concept peculiarly linked to the late nineteenth century. In those years, mathematicians had just laid the foundations for a comprehensive theory of higher-dimensional space. Physicists were beginning to work with the notion of four-dimensional space-time. Philosophers were using the idea of a fourth dimension to solve some of their oldest riddles. And mediums throughout Europe were coming to the conclusion that the spirits of the dead consist of four-dimensional ectoplasm. There was an immense popular interest in the fourth dimension, and *Flatland*, subtitled "A Romance of Many Dimensions," was an immediate success.

Abbott's method was to describe a two-dimensional square's difficulties in imagining a third dimension of space. As we read of A Square's struggles, we become better able to understand our own difficulties in imagining a fourth dimension. The fourth dimension is to us what the third dimension is to the Flatlanders.

This powerful analogy is the rarest of things: a truly new idea. I often used to ask myself where Abbott might have gotten such an idea. When Gray University granted me my sabbatical last year, I determined to go to London and look through Abbott's papers and publications. Could *Flatland* have been inspired by A.F. Moebius's *Barycentric Calculus* of 1827? Might Abbott have corresponded with C. H. Hinton, eccentric author of the 1880 essay, "What is the Fourth Dimension?" Or is *Flatland* nothing more than the inspired reworking of certain ideas in Plato's *Republic*?

Abbott wrote many other books in his lifetime, all crashingly dull: *How to Parse, the Kernel and the Husk—Letters on Spiritual Christianity, English Lessons for English People, A Shakespearian Grammar, Parables for Children,* and so on. Except for *Flatland*, all of Abbott's books are just what one would expect from a Victorian clergyman, headmaster of the City of London School. Where did Abbott find his inspiration for *Flatland*? The answer is stranger than I could ever have imagined.

It was an unnaturally hot day in July. The London papers were full of stories about the heat wave. One man reported that three golf balls had exploded in the heat of his parked car. All the blackboards in a local school had cracked. Numerous pigeons had died and fallen to the sidewalks. I finished my greasy breakfast and set forth from my hotel, an unprepossessing structure not far from St. Paul's Cathedral.

My plan for the day was to visit the site of the old City of London School on Cheapside at Milk Street. Abbott attended the school himself, and then returned as headmaster for the years 1865–1889. Under Abbott's leadership the school moved to a new building in 1882, but I had a feeling that some valuable clue to his psychology might still be found in the older building.

To my disappointment, nothing of the old building remained . . . at least nothing that I could see. Much of Cheapside was destroyed during the Blitz. Flimsy concrete and metal structures have replaced what stood there before. I came to a halt at the corner of Cheapside and Milk, utterly discouraged.

Sweat trickled down my sides. A red double-decker labored past, fouling the heavy air with its exhaust. Ugly, alien music drifted out of the little food shops. I was jostled by men and women of every caste and color: masses of people, hot and impatient, inescapable as the flow of time.

I pushed into a wretched Pakistani snack bar and ordered a beer. They had none. I settled for a Coke. I tried to imagine Edwin Abbott walking through this dingy space one hundred years ago.

The girl behind the counter handed me my Coke. Her skin had a fine coppery color, and her lips were like chocolate ice cream. She didn't smile, but neither did she frown. Desperate with loneliness and disorientation, I struck up a conversation.

"Have you been here long?"

"I was born in London." Her impeccable accent came as a rebuke. "My father owns this shop now for five years."

"Do you know I came all the way from America just to visit this shop?"

She laughed and looked away. A girl in a big city learns to ignore madmen.

"No, no," I insisted. "It's really true. Look . . ." I took out my dog-eared first edition of *Flatland*, this very copy in whose margins I now write. "The man who wrote this book was headmaster of a school that stood near this spot."

"What school?"

"The City School of London. They moved it to the Victoria Embankment in 1882."

"Then you should go there. Here we have only food." For some reason the sight of Abbott's book had caused her cheeks to flush an even darker hue.

"I'll save that trip for another day. Don't you want to know what the book is about?"

"I do know. It is about flat creatures who slide around in a plane."

The readiness of her response astonished me. But before I could pose another question, the girl had turned to serve another customer, a turbaned Sikh with a pockmarked face. I scanned the menu, looking for something else to order.

"Could I have some of the spicy meatballs, please?"

"Certainly."

"What's your name?"

"Deela."

She failed to ask mine, so I volunteered the information. "I'm Bob. Professor Robert Ackley of Gray University."

"And what do you profess?" She set the plate of meatballs down with an encouraging click.

"Mathematics. I study the fourth dimension, just as Abbott did. Have you really read *Flatland*?"

Deela glanced down the counter, as if fearful of being overheard. "I have not *read* it, I . . ."

The Sikh interrupted then, calling for butter on his rice. I sampled one of the meatballs. It was hot and dry as desert sand.

"Could I have another Coke, please?"

"Are you rich?" Deela whispered unexpectedly.

Was she hoping for a date with me? Well, why not? This was the longest conversation I'd had with anyone since coming to London. "I'm well off," I said, hoping to make myself attractive. "I have a good position, and I am unmarried. Would you like to have dinner with me?"

This proposal seemed to surprise Deela. She covered her mouth with one hand and burst into high laughter. Admittedly I am no ladies' man, but this really seemed too rude to bear. I put away my book and rose to my feet.

"What do I owe you?"

"I'm sorry I laughed, Robert. You surprised me. Perhaps I will have dinner with you some day." She lowered her voice and leaned closer. "Downstairs here there is something you should see. I was hoping that you might pay to see it."

It seemed very hot and close in this little restaurant. The inclination of the Sikh's turban indicated that he was listening to our conversation. I had made a fool of myself. It was time to go. Stiffly I paid the bill and left. Only when I stepped out on the street and looked at my change did I realize that Deela had given me a note.

Robert—

Flatland is in the basement of our shop. Come back at closing time and I will show it to you. Please bring one hundred pounds. My father is ill.

Deela.

I turned and started back into the shop. But Deela made a worried face and placed her fingers on her lips. Very well, I could wait. Closing time, I noted, was ten P.M.

I spent the rest of the day in the British Museum, ferreting out obscure books on the fourth dimension. For the first time I was able to hold in my hands a copy of J. K. F. Zöllner's 1878 book, *Transcendental Physics*. Here I read how a spirit from hyperspace would be able to enter a closed room by coming in, not through walls or ceiling, but through the "side" of the room lying open to the fourth dimension.

Four-dimensional spirits . . . long sought, but never found! Smiling a bit at Zöllner's gullibility, I set his book down and reread Deela's note. *Flatland is in the basement of our shop*. What could she mean by this? Had they perhaps found Abbott's original manuscript in the ruined foundations of the old City School? Or did she mean something more literal, something more incredible, something more bizarre than spirits from the fourth dimension?

The whole time in the library, I had the feeling that someone was watching me. When I stepped back onto the street, I realized that I was indeed being followed. It was the Sikh, his obstinate turban always half a block behind me. Finally I lost him by going into a movie theater, leaving by the rear exit, and dashing into the nearest pub.

I passed a bland few hours there, drinking the warm beer and eating the stodgy food. Finally it was ten P.M.

Deela was waiting for me in the darkened shop. She let me in and locked the door behind me.

"Did you bring the money?"

The empty shop felt very private. Deela's breath was spicy and close. What had I really come for?

"Flatland," stated Deela, "is in the basement. Did you bring the money?"

I gave her a fifty-pound note. She flattened it out and held it up to examine it by the street-light. Suddenly there was a rapping at the door. The Sikh!

"Quick!" Deela took me by the arm and rushed me behind the counter and down a narrow hallway. "Down there," she said, indicating a door. "I'll get rid of him." She trotted back out to the front of the shop.

Breathless with fear and excitement, I opened the shabby door and stepped down onto the dark stairs.

The door swung closed behind me, muffling the sound of Deela's voice. She was arguing with the Sikh, though without letting him in. I moved my head this way and that, trying to make out what lay in the basement. Deela's faint voice grew shriller. There was what looked like a ball of light floating at the foot of the stairs. An oddly patterned ball of light some three feet across. I went down a few more steps to have a closer look. The thing was sort of like a huge lens, a lens looking onto . . .

Just then there came the sound of shattering glass. The Sikh had smashed his way in! The clangor of the shop's burglar alarm drowned out Deela's wild screams. Footsteps pounded close by and the door at the head of the stairs flew open.

"Come back up, Professor Ackley," called the Sikh. His voice was high and desperate. "You are in great danger."

But I couldn't tear myself away from the glowing sphere. It appeared to be an Einstein–Rosen bridge, a space tunnel leading into another universe. The other universe seemed to contain only one thing: an endless glowing plane filled with moving forms. Flatland.

The Sikh came clattering down the stairs. My legs made a decision. I leaped forward, through the space tunnel and into another world.

I landed on all fours . . . there was a sort of floor about a yard below the plane of Flatland. When I stood up, it was as if I were standing waist-deep in an endless, shiny lake. My fall through the Flatlanders' space had smashed up one of their houses. Several of them were nosing at my waist, wondering what I was. To my surprise, I could feel their touch quite distinctly. They seemed to have a thickness of several millimeters.

The mouth of the space tunnel was right overhead, a dark sphere framing the Sikh's excited little face. He reached down as if to grab me. I quickly squatted down beneath the plane of Flatland and crawled away across the firm, smooth floor. The hazy, bright space shimmered overhead like an endless soap film, effectively shielding me from the Sikh.

I could hear the sound of more footsteps on the stairs. Deela? There were cries, a gunshot, and then silence. I poked my head back up, being careful not to bump any Flatlanders. The dark opening of the space tunnel was empty. I was safe, safe in Flatland. I rose up to my full height and surveyed the region around me.

I was standing in the middle of a "street," that is to say, in the middle of a clear path lined with Flatland houses on either side. The houses had the form of large squares and rectangles, three to five feet on a side. The Flatlanders themselves were as Abbott has described them: women are short Lines with a bright eye at one end, the soldiers are very sharp isoceles Triangles, and there are Squares, Pentagons, and other Polygons as well. The adults are, on the average, about twelve inches across.

The buildings that lined my street bore signs in the form of strings of colored dots along their outer walls. To my right was the house of a childless Hexagon and his wife. To my left was the home of an equilateral Triangle, proud father of three little Squares. The Triangle's door, a hinged line segment, stood ajar. One of his children, who had been playing in the street, sped inside, frightened by my appearance. The plane of Flatland cut me at the waist and arms, giving me the appearance of a large blob

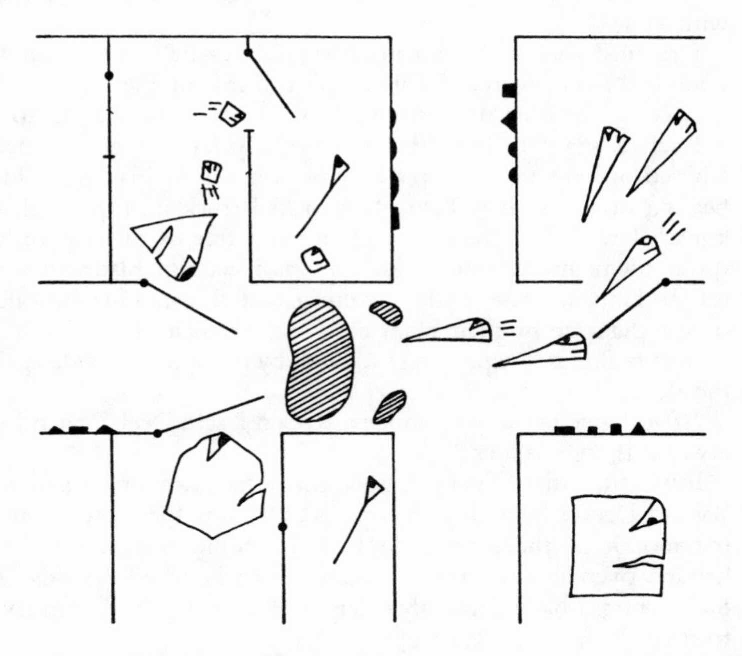

flanked by two smaller blobs—a weird and uncanny spectacle, to be sure.

Now the Triangle stuck his eye out of his door to study me. I could feel his excited voice vibrating the space touching my waist. Flatland seemed to be made of a sort of jelly, perhaps one-sixteenth of an inch thick.

Suddenly I heard Deela calling to me. I looked back at the dark mouth of the tunnel, floating about eight feet above the mysterious ground on which I stood. I walked towards it, staying in the middle of the street. The little line-segment doors slammed as I walked past, and I could look down at the Flatlanders cowering in their homes.

I stopped under the tunnel's mouth and looked up at Deela. She was holding a coiled-up rope ladder.

"Do you want to come out now, Robert?" There was something cold and unpleasant about her voice.

"What happened to the Sikh?"

"He will not bother us again. How much money do you have with you?"

I recalled that so far I had only paid her half of the hundred pounds. "Don't worry, I'll give you the rest of the money." But how could she even think of money with a wonder like this to . . .

I felt a sharp pain in the small of my back, then another. I whirled around to see a platoon of two dozen Flatland soldiers bearing down on me. Two of them had stuck into my back like knives. I wrenched them out, lifted them free of their space, and threw them into the next block. I was bleeding! Blade-thick and tough-skinned, these soldiers were a real threat. One by one, I picked them up by their blunt ends and set them down inside the nearest building. I kept them locked in by propping my side against the door.

"If you give me all your money, Robert," said Deela, "then I will lower this rope ladder."

It was then that I finally grasped the desperation of my situation. Barring Deela's help, there was no possible way for me to get up to the mouth of the tunnel. And Deela would not help unless I handed over all my cash . . . some three hundred pounds. The Sikh, whom I had mistakenly thought of as enemy, had been trying to save me from Deela's trap!

"Come on," she said. "I don't have all night."

There were some more soldiers coming down the street after me. I reached back to feel my wounds. My hand came away wet with blood. It was interesting here, but it was clearly time to leave.

"Very well, you nasty little half-breed thief. Here is all the money I have. Three hundred pounds. The police, I assure you, will hear of this." I drew the bills out and held them up to the tunnel-mouth. Deela reached through, snatched the money, and then disappeared. The new troop of soldiers was almost upon me.

"Hurry!" I shouted. "Hurry up with the ladder! I need medical attention!" Moving quickly, I scooped up the soldiers as they came. One got past my hand and stabbed me in the stomach. I grew angry, and dealt with the remaining soldiers by poking out their hearts.

When I was free to look up at the tunnel-mouth again, I saw a sight to chill the blood. It was the Sikh, eyes glazed in death, his arms dangling down towards me. I realized that Deela had shot him. I grabbed one of his hands and pulled, hoping to lift myself up into the tunnel. But the corpse slid down, crashed through Flatland, and thudded onto the floor at my feet.

"Deela!" I screamed. "For the love of God!"

Her face appeared again . . . but she was no longer holding the rope ladder. In its stead she held a pistol. Of course it would not do to set me free. I would make difficulties. With my body already safe in this dimensional oubliette, it would be nonsense to set me free. Deela aimed her gun.

As before, I ducked below Flatland's opalescent surface and crawled for dear life. Deela didn't even bother shooting.

"Good-bye, Robert," I heard her calling. "Stay away from the tunnel or else!" This was followed by her laughter, her footsteps, the slamming of the cellar door, and then silence.

That was two days ago. My wounds have healed. The Sikh has grown stiff. I made several repellent efforts to use his corpse as a ladder or grappling-hook, but to no avail. The tunnel mouth is too high, and I am constantly distracted by the attacks of the isosceles Triangles.

But my situation is not entirely desperate. The Flatlanders are, I have learned, edible, with a taste something like very moist smoked salmon. It takes quite a few of them to make a meal, but they are plentiful, and they are easy to catch. No matter how tightly they lock their doors, they never know when the five globs of my fingers will appear like Zöllner's spirits to snatch them away.

I have filled the margins of my beloved old *Flatland* now. It is time to move on. Somewhere there may be another tunnel. Before leaving, I will throw this message up through the tunnel mouth. It will lie beneath the basement stairs, and someday someone will find it.

Farewell, reader, and do not pity me. I was but a poor laborer in the vineyard of knowledge—and now I have become the Lord of Flatland.

NORMAN KAGAN

The Mathenauts

It happened on my fifth trip into the spaces, and the first ever made under the private-enterprise acts. It took a long time to get the P.E.A. through Congress for mathenautics, but the precedents went all the way back to the Telstar satellite a hundred years ago, and most of the concepts are in books anyone can buy, though not so readily understand. Besides, it didn't matter if BC-flight was made public or not. All mathenauts are crazy. Everybody knows that.

Take our crew. Johnny Pearl took a pin along whenever he went baby-sitting for the grad students at Berkeley, and three months later the mothers invariably found out they were pregnant again. And Pearl was our physicist.

Then there was Goldwasser. Ed Goldwasser always sits in those pan-on-a-post cigarette holders when we're in New York, and if you ask him, he grumbles; "Well, it's an ashtray, ain't it?" A punster and a pataphysicist. I would never have chosen him to go, except that he and I got the idea together.

Ted Anderson was our metamathematician. He's about half a nanosecond behind Ephraim Cohen (the co-inventor of BC-flight) and has about six nervous breakdowns a month trying to pass him. But he's got the best practical knowledge of the BC-drive outside Princeton—if practical knowledge means anything with respect to a pure mathematical abstraction.

And me—topologist. A topologist is a man who can't tell a doughnut from a cup of coffee. (I'll explain that some other time.) Seriously, I specialize in some of the more abstruse properties of geometric structures. "Did Galois discover that theorem before or after he died?" is a sample of my conversation.

Sure, mathenauts are mathenuts. But as we found out, not quite mathenutty enough.

The ship, the *Albrecht Dold,* was a twelve-googol scout that Ed Goldwasser and I'd picked up cheap from the NYU Courant Institute. She wasn't the Princeton IAS. *Von Neumann,* with googolplex coils and a chapter of the DAR, and she wasn't one of those new toys you've been seeing for a rich man and his grandmother. Her coils were DNA molecules, and the psychosomatics were straight from the Brill Institute at Harvard. A sweet ship. For psychic ecology we'd gotten a bunch of kids from CUNY, commonsense types—business majors, engineers, pre-meds. But kids.

I was looking over Ephraim Cohen's latest paper, *Nymphomaniac Nested Complexes with Rossian Irrelevancies* (old Ice Cream Cohen loves sexy titles), when the trouble started. We'd abstracted, and Goldwasser and Pearl had signaled me from the lab that they were ready for the first tests. I made the *Dold* invariant, and shoved off through one of the passages that linked the isomorphomechanism and the lab. (We kept the ship in free fall for convenience.) I was about halfway along the tube when the immy failed and the walls began to close in.

I spread my legs and braked against the walls of the tube, believing with all my might. On second thought I let the walls sink in and braked with my palms. It would've been no trick to hold the walls for a while. Without the immy my own imagination would hold them, this far from the CUNY kids. But that might've brought more trouble—I'd probably made some silly mistake, and the kids, who might not notice a simple contraction or shear, would crack up under some weirdomorphism. And if we lost the kids . . .

So anyway I just dug my feet in against the mirage and tried to slow up, on a surface that no one'd bothered to think any friction into. Of course, if you've read some of the popular accounts of math-sailing, you'd think I'd just duck back through a hole in the fiftieth dimension to the immy. But it doesn't work out that way. A ship in BC-flight is a very precarious structure in a philosophical sense. That's why we carry a psychic ecology, and that's why Brill conditioning takes six years, plus, with a Ph.D. in pure math, to absorb. Anyway, a mathenaut should never forget his postulates, or he'll find himself floating in 27-space, with nary a notion to be named.

Then the walls really did vanish—NO!—and I found myself at the junction of two passages. The other had a grabline. I caught it and rebounded, then swarmed back along the tube. After ten seconds I was climbing down into a funnel. I caught my breath, swallowed some Dramamine, and burst into the control room.

The heart of the ship was pulsing and throbbing. For a moment I thought I was back in Hawaii with my aqualung, an invader in a shifting, shimmering world of sea fronds and barracuda. But it was no immy, no immy—a rubber room without the notion of distance that we take for granted (technically, a room with topological properties but no metric ones). Instrument racks and chairs and books shrank and ballooned and twisted, and floor and ceiling vibrated with my breath.

It was horrible.

Ted Anderson was hanging in front of the immy, the isomorphomechanism, but he was in no shape to do anything. In fact, he was in no shape at all. His body was pulsing and shaking, so his hands were too big or too small to manipulate the controls, or his eyes shrank or blossomed. Poor Ted's nerves had gone again.

I shoved against the wall and bulleted toward him, a fish in a weaving, shifting undersea landscape, concentrating desperately on my body and the old structure of the room. (This is why physical training is so important.) For an instant I was choking and screaming in a hairy blackness, a nightmare inside-out total inversion; then I was back in the control room, and had shoved Ted away from the instruments, cursing when nothing happened, then bracing against the wall panels and shoving again. He drifted away.

The immy was all right. The twiddles circuits between the CUNY kids and the rest of the *Dold* had been cut out. I set up an orthonormal system and punched the immy.

Across the shuddering, shifting room Ted tried to speak, but found it too difficult. Great Gauss, he was lucky his aorta hadn't contracted to a straw and given him a coronary! I clamped down on my own circulatory system viciously, while he struggled to speak. Finally he kicked off and came tumbling toward me, mouthing and flailing his notebook.

I hit the circuit. The room shifted about and for an instant Ted Anderson hung, ghostly, amid the isomorphomechanism's one-to-ones. Then he disappeared.

The invention of BC-flight was the culmination of a century of work in algebraic topology and experimental psychology. For thousands of years men had speculated as to the nature of the world. For the past five hundred, physics and the physical sciences had held sway. Then Thomas Brill and Ephraim Cohen peeled away another layer of the reality union, and the space sciences came into being.

If you insist on an analogy—well, a scientist touches and probes the real universe, and abstracts an idealization into his head. Mathenautics allows him to grab himself by the scruff of the neck and pull himself up into the idealization. See—I *told* you.

Okay, we'll try it slowly. Science assumes the universe to be ordered, and investigates the nature of the ordering. In the "hard" sciences, mathematics is the basis of the ordering the scientist puts on nature. By the twentieth century, a large portion of the physical processes and materials in the universe were found to submit to such an ordering (e.g.: analytic mechanics and the motions of the planets). Some scientists were even applying mathematical structures to aggregates of living things, and to living processes.

Cohen and Brill asked (in ways far apart), "If order and organization seem to be a natural part of the universe, why can't we remove these qualities from coarse matter and space, and study them separately?" The answer was BC-flight.

Through certain purely mathematical "mechanisms" and special psychological training, selected scientists (the term "mathenaut" came later, slang from the faddy "astronautics") could be shifted into the abstract.

The first mathenautical ships were crewed with young scientists and mathematicians who'd received Tom Brill's treatments and Ephraim Cohen's skull-cracking sessions on the BC-field. The ships went into BC-flight and vanished.

By the theory, the ships didn't *go* anywhere. But the effect was somehow real. Just as a materialist might *see* organic machines instead of people, so the mathenauts saw the raw mathematical structure of space—Riemann space, Hausdorf space, vector space—without matter. A crowd of people existed as an immensely complicated *something* in vector space. The study of these *somethings* was yielding immense amounts of knowledge. Pataphysics, patasociology, patapsychology were wild, baffling new fields of knowledge.

But the math universes were strange, alien. How could you learn to live in Flatland? The wildcat minds of the first crews were too creative. They became disoriented. Hence the immies and their power supplies—SayCows, DaughtAmRevs, the CUNY kids—fatheads, stuffed shirts, personality types that clung to common sense where there was none, and preserved (locally) a ship's psychic ecology. Inside the BC-field, normalcy. Outside, raw imagination.

Johnny, Ted, Goldy, and I had chosen vector spaces with certain topological properties to test Goldy's commercial concept. Outside the BC-field there was dimension but no distance, structure but no shape. Inside—

"By Riemann's tensors!" Pearl cried.

He was at the iris of one of the tubes. A moment later Ed Goldwasser joined him. "What happened to Ted?"

"I—I don't know. No—yes, I do!"

I released the controls I had on my body, and stopped thinking about the room. The immy was working again. "He was doing something with the controls when the twiddles circuit failed. When I got them working again and the room snapped back into shape, he happened to be where the immy had been. The commonsense circuits rejected him."

"So where did he go?" asked Pearl.

"I don't know."

I was sweating. I was thinking of all the things that could've happened when we lost the isomorphomechanism. Some subconscious twitch and you're rotated half a dozen dimensions out of phase, so you're floating in the raw stuff of thought, with maybe a hair-thin line around you to tell you where the ship has been. Or the ship takes the notion to shrink pea-size, so you're squeezed through all the tubes and compartments and smashed to jelly when we orthonormalize. Galois! We'd been lucky.

The last thought gave me a notion. "Could we have shrunk so we're inside his body? Or he grown so we're floating in his liver?"

"No," said Goldy. "Topology is preserved. But I don't—or, hell—I really don't know. If he grew so big he was outside the psychic ecology, he might just have faded away." The big pataphysicist wrinkled up his face inside his beard. "*Alice* should be required reading for mathenauts," he muttered. "The real trouble is no one has ever been outside and been back to tell about it. The animal experiments and the *Norbert Wiener* and Wilbur on the *Paul R. Halmos*. They just disappeared."

"You know," I said, "You can map the volume of a sphere into the whole universe using the ratio IR:R equals R:OR, where IR and OR are the inside and outside distances for the points. Maybe that's what happened to Ted. Maybe he's just outside the ship, filling all space with his metamath and his acne?"

"Down boy," said Goldwasser. "I've got a simpler suggestion. Let's check over the ship, compartment by compartment. Maybe he's in it somewhere, unconscious."

But he wasn't on the ship.

We went over it twice, every tube, every compartment. (In reality, a mathenautic ship looks like a radio, ripped out of its case and flying through the air.) We ended up in the ecology section, a big Broadway-line subway car that roared and rattled in the middle of darkness in the middle of nothing. The CUNY kids were all there—Freddi Urbont clucking happily away to her boyfriend, chubby and smily and an education major; Byron and Burbitt, electonics engineers, ecstatic over the latest copy of *C-Quantum;* Stephen Seidmann, a number-theory major, quietly proving that since Harvard is the best school in the world, and CUNY is better than Harvard, that CUNY is the best school in the world; two citizens with nose jobs and names I'd forgotten, engaged in a filthy discussion of glands and organs and meat. The walls were firm, the straw seats scratchy and uncomfortable. The projectors showed we were just entering the 72nd Street stop. How real, how comforting! I slid the door open to rejoin Johnny and Ed. The subway riders saw me slip into free fall, and glimpsed the emptiness of vector space.

Hell broke loose!

The far side of the car bulged inward, the glass smashing and the metal groaning. The CUNYs had no compensation training!

Freddi Urbont burst into tears. Byron and Burbitt yelled as a bubble in the floor swallowed them. The wall next to the nose jobs sprouted a dozen phallic symbols, while the seat bubbled with breasts. The walls began to melt. Seidmann began to yell about the special status of N. Y. City University Honors Program students.

Pearl acted with a speed and surety I'd never have imagined. He shoved me out of the way and launched himself furiously at the other end of the car, now in free fall. There he pivoted, smiled horribly, and at the top of his lungs began singing "The Purple and the Black."

Goldy and I had enough presence of mind to join him. Concentrating desperately on the shape and form of the car, we blasted the air with our devotion to Sheppard Hall, our love of Convent Avenue, and our eternal devotion to Lewisohn Stadium. Somehow it saved us. The room rumbled and twisted and reformed, and soon the eight of us were back in the tired old subway car that brought its daily catch of Beavers to 137th Street.

The equilibrium was still precarious. I heard Goldwasser telling the nose jobs his terrible monologue about the "Volvo I want to buy. I can be the first to break the door membranes, and when I get my hands on that big, fat steering wheel, ohh!, it'll be a week before I climb out of it!"

Pearl was cooing to Urbont how wonderful she was as the valedictorian at her junior high, how great the teaching profession was, and how useful, and how interesting.

As for me: "Well, I guess you're right, Steve. I should have gone to CUNY instead of Berkeley."

"That's right, Jimmy. After all, CUNY has some of the best number-theory people in the world. And some of the greatest educators, too. Like Dean Cashew who started the Privileged Student Program. It sure is wonderful."

"I guess you're right, Steve."

"I'm right, all right. At schools like Berkeley, you're just another student, but at CUNY you can be a P.S., and get all the good professors and small classes and high grades."

"You're right, Steve."

"I'm right, all right. Listen, we have people that've quit Cornell and Harvard and M.I.T. Of course, they don't do much but run home after school and sit in their houses, but their parents all say how much happier they are—like back in high school . . ."

When the scrap paper and the gum wrappers were up to our knees and there were four false panhandlers in the car, Johnny called a halt. The little psychist smiled and nodded as he walked the three of us carefully out the door.

"Standard technique," he murmured to no one in particular. "Doing *something* immediately rather than the best thing a while later. Their morale was shot, so I—" He trailed off.

"Are they really that sensitive?" Goldwasser asked. "I thought their training was better than that."

"You act like they were components in an electronics rig," said Pearl jerkily. "You know that Premedial Sensory Perception, the ability to perceive the dull routine that normal people ignore, is a very delicate talent!"

Pearl was well launched. "In the dark ages such people were called dullards and subnormals. Only now, in our enlightened age, do we realize their true ability to know things outside the ordinary senses—a talent vital for BC-flight."

The tedium and meaninglessness of life which we rationalize away—

"A ship is more mind than matter, and if you upset that mind—"

He paled suddenly. "I, I think I'd better stay with them," he said. He flung open the door and went back into the coach. Goldwasser and I looked at each other. Pearl was a trained mathenaut, but his specialty was people, not paramath.

"Let's check the lab," I muttered.

Neither of us spoke as we moved toward the lab—slap a wall, pull yourself forward, twist round some instrumentation—the "reaction swim" of a man in free fall. The walls began to quiver again, and I could see Goldy clamp down on his body and memories of this part of the ship. We were nearing the limits of the BC-field. The lab itself, and the experimental apparatus, stuck out into vector space.

"Let's make our tests and go home," I told Goldy.

Neither of us mentioned Ted as we entered the lab.

Remember this was a commercial project. We weren't patasociologists studying abstract groups, or superpurists looking for the first point. We wanted money.

Goldy thought he had a moneymaking scheme for us, but Goldy hasn't been normal since he took Polykarp Kusch's "Kusch of Death" at Columbia, "Electrodimensions and Magnespace." He was going to build four-dimensional molecules.

Go back to Flatland. Imagine a hollow paper pyramid on the surface of that two-dimensional world. To a Flatlander, it is a triangle. Flop down the sides—four triangles. Now put a molecule in each face—one molecule, four molecules. And recall that you have infinite dimensions available. Think of the storage possibilities alone. All the books of the world in a viewer, all the food in the world in your pack. A television the size of a piece of paper; circuits looped through dim-19. Loop an entire industrial plant through hyperspace, and get one the size and shape of a billboard. Shove raw materials in one side—pull finished products out the other!

But how do you make 4-dim molecules? Goldy thought he had a way, and Ted Anderson had checked over the math and pronounced it workable. The notion rested in the middle of the lab: a queer, half-understood machine of mind and matter called Grahm-Schmidt generator.

"Jeez, Ed! This lab looks like your old room back in Diego Borough."

"Yeah," said Goldwasser. "Johnny said it would be a good idea. Orientation against *that*."

That was the outside of the lab, raw topological space, without energy or matter or time. It was the shape and color of what you see in the back of your head.

I looked away.

Goldwasser's room was a duplicate of his old home—the metal desk, the electronic rigs, the immense bookshelves, half filled with physics and half with religious works. I picked up a copy of Stace's *Time and Eternity* and thumbed through it, then put it down, embarrassed.

"Good reading for a place like this." Goldwasser smiled.

He sat down at the desk and began to check out his "instruments" from the locked drawer where he'd kept them. Once he reached across the desk and turned on a tape of Gene Gerard's *Excelsior!* The flat midwestern voice murmured in the background.

"First, I need some hands," said Ed.

Out in the nothingness two pairs of lines met at right angles. For an instant, all space was filled with them, jammed together every which way. Then it just settled down to two.

The lab was in darkness. Goldwasser's big form crouched over the controls. He wore his engineer's boots and his hair long, and a beard as well. He might have been some medieval monk, or primitive witch doctor. He touched a knob and set a widget, and checked in his copy of Birkhoff and MacLane.

"Now," he said, and played with his instruments. Two new vectors rose out of the intersections. "Cross-products. Now I've a right-and a left-handed system."

All the while Gene Gerard was mumbling in the background: "'Ah, now, my pretty,' snarled the Count. 'Come to my bedchamber, or I'll leave you to Igor's mercies.' The misshapen dwarf cackled and rubbed his paws. 'Decide, decide!' cried the Count. His voice was a scream. 'Decide, my dear. *Sex—else, Igor!*'"

"Augh," said Goldwasser, and shut it off. "Now," he said, "I've got some plasma in the next compartment."

"Holy Halmos," I whispered.

Ted Anderson stood beside the generator. He smiled, and went into topological convulsions. I looked away, and presently he came back into shape. "Hard getting used to real space again," he whispered. He looked thinner and paler than ever.

"I haven't got long," he said, "so here it is. You know I was working on Ephraim's theories, looking for a flaw. There isn't any flaw."

"Ted, you're rotating," I cautioned.

He steadied, and continued, "There's no flaw. But the theory is wrong. It's backwards. *This is the real universe,*" he said, and gestured. Beyond the lab topological space remained as always, a blank, the color of the back of your head through your own eyes.

"Now listen to me, Goldy and Johnny and Kidder." I saw that Pearl was standing in the iris of the tube. "What is the nature of intelligence? I guess it's the power to abstract, to conceptualize. I don't know what to say beyond that—I don't know what it is. But I know where it came from! Here! In the math spaces—they're alive with thought, flashing with mind!

"When the twiddles circuits failed, I cracked. I fell apart, lost faith in it all. For I had just found what I thought was a basic error in theory. I died, I vanished . . .

"But I didn't. I'm a metamathematician. An operational philosopher, you might say. I may have gone mad—but I think I passed a threshold of knowledge. I understand . . .

"They're out there. The things we thought we'd invented ourselves. The concepts and the notions and the pure structures—if you could see them . . ."

He looked around the room, desperately. Pearl was rigid against the iris of the tube. Goldy looked at Ted for a moment, then his head darted from side to side. His hands whitened on the controls.

"Jimmy," Ted said.

I didn't know. I moved toward him, across the lab to the edge of topological space, and beyond the psychic ecology. No time, no space, no matter. But how can I say it? How many people can stay awake over a book of modern algebra, and how many of those can understand?

—I saw a set bubbling and whirling, then take purpose and structure to itself and become a group, generate a second unity element, mount itself and become a group, generate a second unity element, mount itself and become a field, ringed by rings. Near it, a mature field, shot through with ideals, threw off a splitting field in a passion of growth, and became complex.

—I saw the life of the matrices; the young ones sporting, adding and multiplying by a constant, the mature ones mating by composition: male and female make male, female and male make female—sex through anticommutativity! I saw them grow old, meeting false identities and loosing rows and columns into nullity.

—I saw a race of vectors, losing their universe to a newer race of tensors that conquered and humbled them.

—I watched the tyranny of the Well Ordering Principle, as a free set was lashed and whipped into structure. I saw a partially ordered set, free and happy, broken before the Axiom of Zemelo.

—I saw the point sets, with their cliques and clubs, infinite numbers of sycophants clustering round a Bolzano-Weierstrass aristocrat—the great compact medieval coverings of infinity with denumerable shires—the conflicts as closed sets created open ones, and the other way round.

—I saw the rigid castes of a society of transformations, orthogonal royalty, inner product gentry, degenerates—where intercomposition set the caste of the lower on the product.

—I saw the proud old cyclic groups, father and son and grandson, generating the generations, rebel and black sheep and hero, following each other endlessly. Close by were the permutation groups, frolicking in a way that seemed like the way you sometimes repeat a sentence endlessly, stressing a different word each time.

There was much I saw that I did not understand, for mathematics is deep, and even a mathenaut must choose his wedge of specialty. But that world of abstractions flamed with a beauty and meaning that chilled the works and worlds of men, so I wept in futility.

Presently we found ourselves back in the lab. I sat beside Ted Anderson and leaned on him, and I did not speak for fear my voice would break.

Anderson talked to Johnny and Ed.

"There was a—a race, here, that grew prideful. It knew the Riemann space, and the vector space, the algebras and the topologies, and yet it was unfulfilled. In some way—oddly like this craft," he murmured, gesturing—"they wove the worlds together, creating the real universe you knew in your youth.

"Yet still it was unsatisfied. Somehow the race yearned so for newness that it surpassed itself, conceiving matter and energy and entropy and creating them.

"And there were laws and properties for these: inertia, speed, potential, quantumization. Perhaps life was an accident. It was not noticed for a long time, and proceeded apace. For the proud race had come to know itself, and saw that the new concepts were . . . flawed." Anderson smiled faintly, and turned to Ed.

"Goldy, remember when we had Berkowitz for algebra?" he asked. "Remember what he said the first day?"

Goldwasser smiled. "Any math majors?"

"Hmm, that's good."

"Any physics majors?"

"Physics majors! You guys are just super engineers!"

"Any chemisty majors?"

"Chemistry major! You'd be better off as a cook!"

Ted finished, "And so on, down to the, ahem, baloney majors."

"He was number-happy," said Ed, smiling.

"No. He was right, in a way." Ted continued. "The race had found its new notions were crudities, simple copies of algebras and geometries past. What it thought was vigor was really sloth and decay.

"It knew how to add and multiply, but it had forgotten what a field was, and what commutativity was. If entropy and time wreaked harm on matter, they did worse by this race. It wasn't interested in expeditions through the fiber bundles; rather it wanted to count apples.

"There was conflict and argument, but it was too late to turn back. The race had already degenerated too far to turn back. Then life was discovered.

"The majority of the race took matter for a bride. Its esthetic and creative powers ruined, it wallowed in passion and pain. Only remnants of reason remained.

"For the rest, return to abstraction was impossible. Time, entropy, had robbed them of their knowledge, their heritage. Yet they still hoped and expended themselves to leave, well, call it a 'seed' of sorts."

"Mathematics?" cried Pearl.

"It explains some things," mused Goldwasser softly. "Why abstract mathematics, developed in the mind, turns out fifty years or a century later to accurately describe the physical universe. Tensor calculus and relativity, for example. If you look at it this way, the math was there first."

"Yes, yes, yes. Mathematicians talked about their subject as an art form. One system is more 'elegant' than another if its logical structure is more austere. But Occam's Razor, the law of simplest hypothesis, isn't logical.

"Many of the great mathematicians did their greatest work as children and youths before they were dissipated by the sensual world. In a trivial sense, scientists and mathematicians most of all are described as 'unworldly' . . ."

Anderson bobbled his head in the old familiar way. "You have almost returned," he said quietly. "This ship is really a heuristic device, an aid to perception. You are on the threshold. You have come all the way back."

The metamathematician took his notebook, and seemed to set all his will upon it. "See Ephraim gets this," he murmured. "He, you, I . . . the oneness—"

Abruptly he disappeared. The notebook fell to the floor.

I took it up. Neither Ed nor Johnny Pearl met my eyes. We may have sat and stood there for several hours, numbed, silent. Presently the two began setting up the isomorphomechanism for realization. I joined them.

The National Mathenautics and Hyperspace Administration had jurisdiction over civilian flights then, even as it does today. Ted was pretty important, it seemed. Our preliminary debriefing won us a maximum-security session with their research chief.

Perhaps, as I'd thought passionately for an instant, I'd have done better to smash the immy, rupture the psychic ecology, let the eggshell be shattered at last. But that's not the way of it. For all of our progress, some rules of scientific investigation don't change. Our first duty was to report back. Better heads than ours would decide what to do next.

They did. Ephraim Cohen didn't say anything after he heard us out and looked at Ted's notebook. Old Ice Cream sat there, a big teddy-bear-shaped genius with thick black hair and a dumb smile, and grinned at us. It was in Institute code.

The CUNY kids hadn't seen anything, of course. So nobody talked.

Johnny Pearl married a girl named Judy Shatz and they had fifteen kids. I guess that showed Johnny's views on the matter of matter.

Ed Goldwasser got religion. Zen-Judaism is pretty orthodox these days, yet somehow he found it suited him. But he didn't forget what happened back out in space. His book, *The Cosmic Mind,* came out last month, and it's a good summation of Ted's ideas, with a minimum of spiritual overtones.

Myself. Well, a mathematician, especially a topologist, is useless after thirty, the way progress is going along these days. But *Dim-Dustries* is a commercial enterprise, and I guess I'm good for twenty years more as a businessman.

Goldwasser's Grahm-Schmidt generator worked, but that was just the beginning. Dimensional extensions made Earth a paradise, with housing hidden in the probabilities and automated industries tucked away in the dimensions.

The biggest boon was something no one anticipated. A space of infinite dimensions solves all the basic problems of modern computer-circuit design. Now all components can be linked with short electron paths, no matter how big and complex the device.

There have been any number of other benefits. The space hospitals, for example, where topological surgery can cure the most terrible wounds—and topological psychiatry the most baffling syndromes. (Four years of math is required for pre-meds these days.) Patapsychology and patasociology finally made some progress, so that political and economic woes have declined—thanks, too, to the spaces, which have drained off a good deal of poor

Earth's overpopulation. There are even space resorts, or so I'm told—I don't get away much.

I've struck it lucky. Fantastically so.

The Private Enterprise Acts had just been passed, you'll recall, and I had decided I didn't want to go spacing again. With the training required for the subject, I guess I was the only qualified man who had a peddler's pack, too. Jaffee, one of my friends down at Securities and Exchange, went so far as to say that *Dim-Dustries* was a hyperspherical trust (math is required for pre-laws too). But I placated him and I got some of my mathemateers to realign the Street on a Moebius strip, so he had to side with me.

Me, I'll stick to the Earth. The "real" planet is a garden spot now, and the girls are very lovely.

Ted Anderson was recorded lost in topological space. He wasn't the first, and he was far from the last. Twiddles circuits have burned out, Naught-Ams-Revs have gone mad, and no doubt there have been some believers who have sought out the Great Race.